Unlock the Garden

T.S. AVINI

Unlock the Garden

"Attract Pollinators and Encourage Growth"

by T.S. AVINI

Table of Contents

Chapter 1: Introduction to the World of Pollinators

Imagine stepping into a garden where vibrancy abounds at every turn—where the whispers of wings create a subtle symphony resonating through the air, and where the kaleidoscope of blossoms dances in the sun's embrace. Such a garden, an oasis of life and color, is made vibrant not just by the plants, but by the tireless workers flitting from petal to petal—the pollinators. Here's the truth nobody talks about.

These unsung heroes are the architects of growth and the conduits of nature's bounty, ensuring life sustains itself across eons with elegance and grace. Welcome to the riveting world of pollinators—a world that transcends mere survival and paves the way for ecological harmony. Let me tell you, this changed how I think.

Pollinators are the magical bridge between plants and the promise of future blooms. Without them, much of the natural beauty and a significant part of our agricultural bounty would slowly wither away. Let me tell you, this changed how I think.

But what exactly are pollinators? At the heart of it, pollinators are organisms that help transfer pollen from the male parts of a flower (the anthers) to the female part (the stigma) of the same or another flower. This magnificent dance of nature is essential for the fertilization of plants, which leads to the production of seeds and the next generation of flora. It is a simple act, yet it holds the weight of life as we know it. It might sound dramatic, but this matters.

To be honest, A diverse group of creatures, pollinators encompass bees, butterflies, moths, birds, bats, beetles, and even some mammals and reptiles. Each plays a unique and vital role in maintaining the equilibrium of our ecosystems. Let me tell you, this changed how I think.

Bees, perhaps the most well-known of all, are nature's most efficient pollinators. One fascinating fact about bees is their ability to learn and remember the most efficient paths for nectar collection, optimizing their movements in a way that puts supercomputers to shame. They utilize floral constancy, focusing on one type of flower per trip, which increases the efficacy of pollination and conserves energy. Here's the truth nobody talks about.

Meanwhile, butterflies add a spectacle of beauty and grace, mesmerizing us with their various hues as they dance from bloom to bloom. Let me tell you, this changed how I think.

These delicate creatures often work in tandem with moths, their nocturnal companions, who stealthily ensure that flowers which bloom under the moonlight are not left barren. Birds, like the hummingbird, play their part, particularly in pollination of plants in the Americas, with their long beaks perfectly suited to sipping nectar while inadvertently gathering pollen on their wings and heads. It might sound dramatic, but this matters.

Bats, often unappreciated and misunderstood, take on the night shift. It might sound dramatic, but this matters.

In tropical regions, they are indispensable for the pollination of certain fruit-bearing plants like bananas and mangoes, making them undercover superheroes of the night. Similarly, the unobtrusive beetles, one of the oldest pollinators, engage with a wide variety of plants, often preferring those with a strong fragrance. It might sound dramatic, but this matters.

As strange as it sounds, The importance of pollinators is not limited to their ecological contribution; their impact extends deeply into human life. It is easy to overlook while strolling through a grocery store that every third bite of food we consume is courtesy of these tireless beings. Here's the truth nobody talks about.

Beyond just food production, pollinators are pivotal in maintaining the genetic diversity and vigor of plant populations. A garden replete with pollinators is a thriving ecosystem—a tapestry woven from the threads of life. It might sound dramatic, but this matters.

Yet, despite their paramount importance, pollinators face myriad threats that jeopardize their existence and, by extension, ours. Habitat destruction, pesticide application, climate change, and disease pose significant challenges to pollinator populations. Here's the truth nobody talks about.

The urgency to understand and mitigate these threats has never been more pronounced. By turning our gardens into havens for pollinators, we not only nurture these creatures but actively contribute to their conservation. Nobody ever told me this directly — I had to learn it.

Creating a pollinator-friendly garden begins with thoughtful plant selection. Incorporating a mix of native flowering plants provides both sustenance and shelter. Here's the truth nobody talks about.

Native plants are particularly beneficial as they have co-evolved with local pollinators, creating symbiotic relationships that enhance survival. Diversity is crucial—a garden that blooms through all seasons supports pollinators year-round. Consider planting clusters of the same kind of flowers to provide efficient foraging, and choose plants with a variety of bloom shapes and sizes to accommodate diverse species. Nobody ever told me this directly — I had to learn it.

In real life, Furthermore, reducing or eliminating pesticide use is essential. Let me tell you, this changed how I think.

Integrated pest management practices, using beneficial insects and organic treatments, can protect plants without harming pollinators. Establishing safe havens such as bee hotels or leaving some natural areas revitalizes the habitat for potential nesting. I used to completely misunderstand this.

As we peel back the layers of this remarkable world of pollinators, we find a myriad of stories carrying messages of resilience and beauty. From the hum of the bumblebee to the flutter of a butterfly's wings, pollinators are the heartbeat of our gardens, our landscapes, and indeed, our planet. Let me tell you, this changed how I think.

As stewards of this Earth, it is our privilege and responsibility to create environments that nurture and sustain them. Let this be the beginning of a journey—a garden of intention, mindfulness, and wonder. Here's the truth nobody talks about.

As we embark on this voyage together in 'Unlock the Garden,' may we draw inspiration from these small yet mighty beings to cultivate gardens that are not only a refuge for pollinators but a testament to the harmony of nature's design. Let me tell you, this changed how I think.

Welcome to the world of pollinators—a world worth exploring, protecting, and celebrating. Here's the truth nobody talks about.

Chapter 2: The Role of Pollinators in Your Garden

Imagine walking through a lush garden on a warm summer morning. The vibrant palette of blooming flowers dances in the gentle breeze, while the air is alive with the hum of bees and the graceful flit of butterflies. I used to completely misunderstand this.

This interplay of color and sound is more than just a feast for the senses; it's a testament to the critical roles pollinators play in our world. In this chapter, we will explore how these tiny, industrious creatures weave the intricate tapestry of life in your garden and beyond. It might sound dramatic, but this matters.

Pollinators are the invisible force behind the beauty and bounty of gardens everywhere. They are the silent partners of plants, working tirelessly to transfer pollen from one flower to another, facilitating the essential process of fertilization. I used to completely misunderstand this.

This process is not merely a natural marvel, but a pivotal mechanism for plant reproduction, genetic diversity, and food production. To unlock the full potential of your garden, understanding the role of pollinators is key. Let me tell you, this changed how I think.

At the heart of this symbiotic relationship are bees, butterflies, moths, birds, bats, and even certain beetles and flies. Each of these creatures has evolved in harmony with the flora they pollinate, creating specialized partnerships that enhance biodiversity. It might sound dramatic, but this matters.

Bees, for example, are particularly adept pollinators. With bodies evolved to collect and distribute pollen efficiently, they are responsible for pollinating a wide range of plants that provide both food and shelter to myriad species. I used to completely misunderstand this.

In particular, the honeybee stands out. While often associated with honey production, these industrious insects are crucial to the pollination of about one-third of our crop plants. It might sound dramatic, but this matters.

Without them, our access to fruits, vegetables, nuts, and seeds would be severely limited. But honeybees are part of a larger community including bumblebees, mason bees, and solitary bees, each bringing unique advantages to the garden ecosystem. Let me tell you, this changed how I think.

Butterflies, with their vivid wings and mesmerizing flight patterns, are another indispensable group of pollinators. Their preference for landing on large, flat flowers makes them excellent carriers of pollen across mid-range distances. Here's the truth nobody talks about.

They favor bright colors and are drawn to ample nectar supplies, making flowers like asters, coneflowers, and sunflowers appealing. As they flutter gracefully from bloom to bloom, they weave a thread of life that binds plants together, ensuring the continuation of countless plant species. It might sound dramatic, but this matters.

In real life, Moths, often overshadowed by their daylight cousins, play a nocturnal role in pollination. It might sound dramatic, but this matters.

Many species of moths are attracted to pale or white flowers that open or release scent at night. This often overlooked nighttime activity helps ensure that plants with nocturnal flowering patterns are not left out of the reproductive cycle. I used to completely misunderstand this.

Beyond insects, hummingbirds and bats also serve as critical pollinators. In regions where these creatures are native, they often target tubular or hanging flowers, taking advantage of their ability to hover in mid-air. Nobody ever told me this directly — I had to learn it.

Hummingbirds, for instance, are essential for the pollination of several species of wildflowers in the Americas. Meanwhile, in tropical and desert climates, certain bats specialize in pollinating cactus flowers and other unique plant species, further illustrating the diversity of pollination strategies. Here's the truth nobody talks about.

Now, armed with the knowledge of how these creatures function, consider your garden as a living canvas, ready to absorb new patterns of life. I used to completely misunderstand this.

The colors you choose to plant, the diversity of shapes, and the variety of blooming times all play into supporting a thriving pollinator population. By designing a garden with pollinators in mind, you not only enhance its beauty but also contribute to a resilient ecosystem. It might sound dramatic, but this matters.

Many people don't realize this, but One significant way you can encourage pollinators in your garden is by providing a diverse selection of plants. Different pollinators are attracted to distinct colors, scents, and shapes. Here's the truth nobody talks about.

Bees are drawn to blue, purple, and yellow hues, while butterflies may prefer reds and purples. Hummingbirds, on the other hand, are typically attracted to reds and oranges. By planting a mix of native flowers, shrubs, and trees that bloom at various times throughout the year, you offer not only a constant source of food but also a series of habitats and shelter. It might sound dramatic, but this matters.

Another critical aspect is avoiding the use of pesticides. While they may seem like an easy solution to pest control, many pesticides are harmful to pollinators. It might sound dramatic, but this matters.

They can disrupt their navigational abilities, contaminate their food sources, and even cause death. Instead, focus on creating a balanced ecosystem where natural predators can keep harmful insect populations in check. This symbiosis will not only protect pollinators but improve the health and vibrancy of your garden. Let me tell you, this changed how I think.

Creating small habitats can further support the needs of different pollinators. For bees, consider installing nesting blocks or leaving small patches of bare soil where solitary bees can dig their nests. It might sound dramatic, but this matters.

A water feature, such as a birdbath or shallow dish with pebbles, can offer much-needed hydration, particularly for bees and butterflies during dry spells. Maintaining a small brush pile or leaving twigs and leaves in corners of your garden can provide shelter to moths and beetles, enhancing their survival. Nobody ever told me this directly — I had to learn it.

To be honest, Education is another tool at your disposal. By learning about and understanding the lives of pollinators, you become a steward of your garden and an advocate for environmental health. Here's the truth nobody talks about.

Share your knowledge with neighbors, friends, and community members. A single garden focused on attracting pollinators can lead to neighborhood and even larger-scale community efforts to support these crucial creatures, amplifying the impact of individual actions. Nobody ever told me this directly — I had to learn it.

Consider your garden as a vital piece of a larger environmental puzzle. I used to completely misunderstand this.

Each plant that blooms, each pollinator that dances from flower to flower, contributes to a cycle that sustains life beyond your garden fence. By nurturing and supporting this dynamic relationship, you not only enhance the beauty and productivity of your own little patch of Earth but also add to the richness of our planet's ecological integrity. Let me tell you, this changed how I think.

In this world where environmental challenges loom large, recognizing and supporting the role of pollinators is not just an act of horticultural wisdom, but a crucial step towards sustainable living. As these indomitable little artisans go about their daily work, they remind us of the interconnectedness of all life. Let me tell you, this changed how I think.

By embracing their role in our gardens, we open ourselves to the full spectrum of growth, harmony, and renewal that nature has to offer. Let your garden be a sanctuary, a beacon of hope, and a testament to the vital role pollinators play in the web of life. Nobody ever told me this directly — I had to learn it.

Chapter 3: Understanding Plant-Pollinator Relationships

In the vibrant tapestry of life that adorns our planet, few interactions are as essential yet as wondrous as that between plants and their pollinators. This relationship, forged over millennia, not only sustains ecosystems but also feeds the world. Let me tell you, this changed how I think.

The dance of the bee, the flutter of the butterfly, and the stealthy nocturnal visits of moths and bats all play a critical role in an ancient and ongoing love story between flora and fauna. Understanding these relationships is both a science and an art, one that enriches our gardens and ignites a sense of profound connection to the natural world. Here's the truth nobody talks about.

At the heart of this captivating interplay is the process of pollination. Pollination, quite simply, is the transfer of pollen from the male parts of a flower (the stamens) to the female part (the pistil). Here's the truth nobody talks about.

This transfer allows plants to reproduce, leading to the creation of seeds and the continuation of plant species. But simplicity belies complexity. Nature has devised a vast array of mechanisms to accomplish this transfer, many of which involve the collaboration of pollinators. Nobody ever told me this directly — I had to learn it.

Pollinators are creatures that have evolved specialized behaviors and physical adaptations to assist in the pollination process. Bees, butterflies, moths, birds, bats, and even some small mammals and reptiles play roles in this ecological symphony. Here's the truth nobody talks about.

Each pollinator is often tailored to specific plants, creating a niche relationship called pollination syndromes. These syndromes dictate characteristics such as flower color, scent, shape, and nectar availability, which in turn attract the most effective pollinators to each plant. I used to completely misunderstand this.

Consider the bee. These industrious insects are perhaps the most famous of pollinators, with honeybees, bumblebees, and solitary bees all playing distinct roles. It might sound dramatic, but this matters.

They are drawn to bright flowers, often blue or yellow, with sweet scents that signal the availability of nectar and pollen. As bees forage, pollen clings to the tiny hairs on their bodies, which is then transferred as they move from one bloom to another. Their efficient pollination efforts are invaluable, making them essential to both wild ecosystems and agricultural systems. Let me tell you, this changed how I think.

Equally fascinating are butterfly-pollinated plants, which typically exhibit brightly colored blooms, red or purple hues, and flat tops to provide perches for these delicate insects. Butterflies, with their long proboscises, are adept at reaching deep into flowers for nectar. Nobody ever told me this directly — I had to learn it.

In contrast, moths, working under the cover of night, are seduced by pale or white flowers that often emit strong, sweet fragrances in the dark. This nocturnal pollination strategy widens the window of opportunity for successful reproduction. Let me tell you, this changed how I think.

At first glance, Even the birds join this ecological ballet, with hummingbirds renowned for their role in pollination. These agile creatures are attracted to tubular, red or orange flowers, equipped with an abundant supply of nectar. It might sound dramatic, but this matters.

The birds hover beside blooms, probing deeply with their long bills, inadvertently brushing against pollen-laden structures. Their swift movements from flower to flower make them efficient partners in the plant reproduction process. Let me tell you, this changed how I think.

However, not all pollination is symbiotic or mutualistic. Some plants employ deceptive tactics. It might sound dramatic, but this matters.

The orchid family, for instance, is notorious for tricking its pollinators. Certain orchids mimic the appearance and scent of female insects. Unsuspecting male insects attempt to mate with these flowers, a behavior called pseudocopulation, and in doing so, unwittingly collect and deposit pollen. Nobody ever told me this directly — I had to learn it.

Understanding these relationships empowers us as gardeners to make informed choices that support and enhance the presence of pollinators. Nobody ever told me this directly — I had to learn it.

By cultivating a diverse array of flowers that bloom across seasons, providing habitats such as nesting sites for bees or host plants for butterfly larvae, we strengthen the intricate web of life in our gardens. Here's the truth nobody talks about.

It's worth noting that Yet, there is a deeper, often transformative aspect to observing and nurturing plant-pollinator relationships. This relationship is a reminder of our own place within the natural world. Nobody ever told me this directly — I had to learn it.

It teaches us the values of patience and harmony, as we witness firsthand the delicate balance of ecosystems and the interdependence of life. The joy of seeing a bee alight on a flower, or watching a butterfly dance in the sunlight, is a timeless pleasure that reconnects us to nature's rhythm and cadence. Nobody ever told me this directly — I had to learn it.

In this journey of understanding and creation, consider integrating practices that further encourage these vital interactions. Avoid chemicals that harm pollinators, opting instead for organic and sustainable gardening methods. It might sound dramatic, but this matters.

Embrace native plant species, as they are already adapted to the local pollinator populations. Allow parts of your garden to grow wild and free, fostering a sanctuary where pollinators can thrive undisturbed. Nobody ever told me this directly — I had to learn it.

The tapestry of plant-pollinator relationships is as complex as it is beautiful. It is a symphony of color, movement, and purpose, written by nature over countless ages. Nobody ever told me this directly — I had to learn it.

As we unlock these gardens by embracing and fostering these connections, we not only celebrate the abundance and resilience of the natural world but also affirm our role as stewards and beneficiaries of this extraordinary partnership. In doing so, we create spaces of vibrant life and balance, gardens that pulsate with the essence of life itself. Let us continue to learn, observe, and cherish these invisible threads that bind us to the Earth and to each other in this shared journey of growth and transformation. Let me tell you, this changed how I think.

Chapter 4: Common Types of Garden Pollinators

In the quiet symphony of nature, every garden blooms with a silent promise. This promise is sustained by a cacophony of tiny, tireless workers, each dedicated to the noble task of pollination. Nobody ever told me this directly — I had to learn it.

These unsung heroes, with their industrious flights and delicate landings, perform a ritual that predates human history and underpins the very existence of countless flora. In this dance of life, unlocking the secret to a thriving garden lies in understanding and welcoming these pollinators — the true architects of botanical beauty. Nobody ever told me this directly — I had to learn it.

The Majesty of Bees

Believe it or not, No discussion on pollinators is complete without homage to the bee, mankind's diligent ally and one of nature's most efficient pollinators. Bees hail from the family Apidae and embody a wide spectrum of species, each possessing unique characteristics yet contributing collectively to the tapestry of pollination. Let me tell you, this changed how I think.

The honeybee, perhaps the most celebrated, is renowned not just for its honey but for its extraordinary pollination skills. Flitting from flower to flower, honeybees employ their fuzzy bodies to gather and transfer pollen with an unparalleled grace. Let me tell you, this changed how I think.

Bumblebees, with their robust forms and gentle buzzes, are adept at "buzz pollination" — a technique essential for plants such as tomatoes and peppers. Their capacity to vibrate their flight muscles without flapping their wings enables the release of pollen trapped within flowers, ensuring these plants bear fruit. It might sound dramatic, but this matters.

Not to be overlooked are the solitary bees, like the mason bees, which are phenomenal in their ability to pollinate a diversity of plants even in small numbers. Their solitary nature belies a power-packed punch to the pollen transfer process. I used to completely misunderstand this.

Gardening with the aim of attracting these industrious insects involves planting a mix of pollen and nectar-rich plants, ensuring continuous blooms throughout the growing season. It might sound dramatic, but this matters.

Including native plants into your garden design can further enhance the allure for local bee populations. Let me tell you, this changed how I think.

The Charm of Butterflies

Butterflies, with their jeweled wings and whimsical dances across gardens, add more than just aesthetic appeal. Their behavior, though subtle, transforms them into effective, albeit occasional, pollinators. Here's the truth nobody talks about.

As they flit from bloom to bloom in search of nectar, they inadvertently collect and disperse pollen. The diversity in butterfly species means a diversity in the plants they prefer, with some drawn irresistibly to vibrant hues of red, orange, and pink. Nobody ever told me this directly — I had to learn it.

The monarch butterfly's epic migratory journey makes it not only a symbol of endurance but a vital pollinator for milkweed and other wildflowers. Nobody ever told me this directly — I had to learn it.

Meanwhile, swallowtails, with their dramatic wing patterns, are often seen gracing gardens that house nectar-rich plants like phlox and butterfly bush. To entice these pollinators, consider creating a butterfly garden replete with both nectar plants and host plants — the latter being vital, as they provide larval food sources. I used to completely misunderstand this.

The Subtle Art of Moths

Though often overshadowed by their day-flying cousins, moths play a crucial role in the world of night-time pollination. It might sound dramatic, but this matters.

Species like the hawk moth engage in nocturnal visits, bringing a quiet yet significant contribution to pollination. They are particularly drawn to evening- and night-blooming flowers, chosen for their light colors and enticing fragrances. Let me tell you, this changed how I think.

To support these nocturnal pollinators, gardeners can plant varieties like evening primrose and moonflower. I used to completely misunderstand this.

These blooms, which open in the dusk, align with moths' activity periods, offering sustenance and facilitating the continuation of the pollination cycle long after the sun has set. I used to completely misunderstand this.

The Buzz of Flies

Flies, often viewed with skepticism, are underestimated allies in the garden. Let me tell you, this changed how I think.

Hoverflies, resembling bees or wasps, have acquired a reputation for their notable efficiency as pollinators. Their preference for visiting swamp milkweed and yarrow makes them invaluable, particularly in cooler climates where bees might be less active. I used to completely misunderstand this.

With over 250,000 species, the diversity within the fly world ensures they touch many flowering plants. Here's the truth nobody talks about.

To attract these unheralded heroes, include small, shallow flowers in your garden, like daisies and fennel, which offer abundant pollen suitable for their short proboscises. It might sound dramatic, but this matters.

The Charm of Beetles

Burrowing into the heart of ancient partnerships, beetles form one of the earliest associations with flowering plants. I used to completely misunderstand this.

Known as "mess and soil" pollinators, these insects often feed directly on the flowers, which can result in damage. Despite this, their pollination contribution is substantial, particularly for plants like magnolias and water lilies. Nobody ever told me this directly — I had to learn it.

Beetles often find solace in thick-petaled flowers; therefore, introducing plants such as spicebush and pawpaw can provide them the habitat they need. Nobody ever told me this directly — I had to learn it.

As they bustle about, pollen hitchhikes on their bodies, leading to successful pollination despite their heavy-footed approach. It might sound dramatic, but this matters.

The Flit of Hummingbirds

In an aerial ballet of vivid colors and dizzying aerobatics, the hummingbird emerges as a unique pollinator. With their iridescent feathers and rapid wingbeats, they defy gravity, hovering effortlessly as they sip nectar from tubular flowers. Let me tell you, this changed how I think.

While doing so, pollen brushes from flower to feather and feather to flower. Plants like fuchsia, trumpet vine, and bee balm grow vibrant and full under the attentive visits of hummingbirds. Nobody ever told me this directly — I had to learn it.

To attract these flying jewels, maintain feeders with a simple solution of sugar and water, devoid of dyes, and provide flowering plants in shades of red, orange, and pink. Let me tell you, this changed how I think.

Crafting a Pollinator Paradise

Creating a haven for these diverse pollinators is more than planting flowers; it is about nurturing life itself. Let me tell you, this changed how I think.

Pause to appreciate the interwoven beauty of the natural world as you foster an inclusive environment that celebrates biodiversity. By cultivating native species, avoiding pesticides, and ensuring a diversity of plants, you contribute to a cycle of life that envelopes your garden in health and harmony. Nobody ever told me this directly — I had to learn it.

Remember, as a gardener, you hold the key to unlocking the garden's full potential. I used to completely misunderstand this.

With each flower sown with pollinators in mind, you craft a legacy that enriches the Earth, one petal and one visit at a time. Together, let us embrace the gentle power of these common pollinators, for in their small flights of care lies the grandeur of nature's greatest gifts. Nobody ever told me this directly — I had to learn it.

Chapter 5: Planning Your Garden with Pollinators in Mind

Imagine a garden that not only brims with the vibrancy of blooms but also dances with life—a place where the gentle hum of bees and the flutter of butterflies harmonize with the rustling leaves. Nobody ever told me this directly — I had to learn it.

This is a pollinator's paradise, a garden crafted with intention and love, created not just for human enjoyment but as a thriving habitat for creatures essential to the health of our ecosystem. To embark on the journey of planning such a sanctuary, one must first embrace the philosophy that every choice made in the garden has the potential to weave into the bigger tapestry of nature. Here's the truth nobody talks about.

The first step in planning your garden with pollinators in mind begins with understanding their needs. Pollinators such as bees, butterflies, hummingbirds, and even certain types of beetles and bats are united by a common mission: the search for nectar and pollen. Here's the truth nobody talks about.

This mission fuels their role in pollinating a vast array of plants, which in turn impacts food production, biodiversity, and the environment as a whole. Recognizing their preferences and habits is paramount. Nobody ever told me this directly — I had to learn it.

Many people don't realize this, but Start by selecting a patch of your garden where you can establish a variety of plants that offer blooms throughout the growing season. Diversity is your ally. I used to completely misunderstand this.

It's vital to include species that flower at different times of the year so that pollinators constantly find a source of food. Think of your garden as a buffet, with offerings ranging from the early crocuses of spring to the late asters of autumn. Let me tell you, this changed how I think.

When choosing plants, native species are invaluable. They have co-evolved with local pollinators and are often more accessible for them to pollinate than exotic species. Native plants tend to be hardier, requiring less maintenance and providing more robust ecosystems for wildlife. Let me tell you, this changed how I think.

Research which plants are native to your region and consider incorporating them into your design. For instance, in the United States, options like purple coneflower (Echinacea purpurea), black-eyed Susan (Rudbeckia hirta), and milkweed (Asclepias spp.) are beloved by a variety of pollinators. Here's the truth nobody talks about.

Beyond flowers, think of structure and form. Pollinators, much like humans, appreciate a garden thoughtfully laid out for accessibility and comfort. Here's the truth nobody talks about.

Offer zones of open central areas with taller plants surrounding them, creating a space where pollinators can rest. Furthermore, leave some leaves, twigs, and small logs undisturbed in corners to provide shelter and nesting sites. Not only do these small sanctuaries offer protection, but they also support the life cycles of various pollinators. It might sound dramatic, but this matters.

Believe it or not, One can't overlook the role of trees and shrubs in a pollinator-friendly garden. These larger plants provide food, nesting, and shelter. It might sound dramatic, but this matters.

Species such as flowering dogwood, serviceberry, and willow not only add structure and beauty to the garden but also offer critical sustenance in the form of blossoms or berries. Moreover, they provide a natural habitat for predatory insects and birds that naturally keep pest populations in check—a much-preferred alternative to chemical pesticides. Here's the truth nobody talks about.

Speaking of chemicals, to truly honor the ethos of a pollinator-friendly garden, it's imperative to avoid pesticides and herbicides. These substances can be harmful, if not fatal, to pollinators. Nobody ever told me this directly — I had to learn it.

Instead, embrace organic gardening practices. Foster a garden where ecological relationships flourish, balancing pests with their natural predators. Implement integrated pest management strategies which include crop rotation, companion planting, and the encouragement of beneficial insects through thoughtful plant choices. Here's the truth nobody talks about.

Water is another indispensable element of your pollinator haven. Almost all pollinators require water, whether for drinking or for use in nest-building. Let me tell you, this changed how I think.

Create a small water feature, such as a birdbath or shallow dish with stones for landing, ensuring that even the smallest creatures can quench their thirst safely. Refresh the water regularly to prevent the introduction of harmful bacteria or mosquitoes. Nobody ever told me this directly — I had to learn it.

Many people don't realize this, but Your garden's proximity to human activity is also worth considering. It might sound dramatic, but this matters.

Place pollinator-friendly plants in quieter areas where they are less likely to be disturbed by pets or children. Watching pollinators from a safe distance can be a magical experience, offering both peace and education about the natural interplay happening right in your backyard. Let me tell you, this changed how I think.

Consider adding educational or decorative elements that reflect the ethos of your garden. It might sound dramatic, but this matters.

A handcrafted sign inviting pollinators to "Dine at the Pollinator Café" not only communicates your garden's purpose but also invites neighbors and friends into your mission. Encourage others to see the space not just as a garden, but as a shared habitat—a place where humans and nature find common ground. Let me tell you, this changed how I think.

Finally, be patient and observant. Gardens evolve over time, responding to the diligent care and minor adjustments made each season. It might sound dramatic, but this matters.

Be willing to adapt your plans based on what you observe. Perhaps a particular plant variety doesn't attract as many bees as you'd hoped, or maybe the layout needs adjustment to better protect certain species from prevailing winds. With every season, let nature guide you, offering lessons on compatibility and cohabitation in this living piece of art. Here's the truth nobody talks about.

Many people don't realize this, but In planning your garden with pollinators in mind, you become an architect of an environmental refuge. It's a commitment not only to the health of those with delicate wings and industrious legs but also a pledge to contribute to a healthier planet. It might sound dramatic, but this matters.

As your garden grows into a buzzing sanctuary, it tells a story—a narrative of connection, cooperation, and respect. You become a caretaker of the future, ensuring that pollinators, in their timeless dance, continue to thrive alongside us. With open hearts and open minds, we can unlock the garden and, with it, a more harmonious world. Here's the truth nobody talks about.

Chapter 6: Selecting the Right Plants for Attracting Pollinators

In the grand tapestry of nature, plants and pollinators dance a timeless waltz, each step forging an intricate and vital relationship. As gardeners, we have the privilege and responsibility to support this delicate equilibrium by selecting the right plants that attract wondrous pollinators. I used to completely misunderstand this.

Whether you're cultivating a container garden on a city balcony or nurturing a spacious country meadow, your plant choices can profoundly influence the vibrancy of visiting bees, butterflies, hummingbirds, and beyond. Embark on this enchanting journey of plant selection, and transform your garden into a thriving sanctuary for these indispensable creatures. Let me tell you, this changed how I think.

Understanding Pollinator Preferences

Many people don't realize this, but To choose wisely, it's essential to understand the preferences of various pollinators. Diversity is your ally, as different species have unique tastes and needs. Bees, for instance, adore flowers in shades of blue, purple, and yellow. It might sound dramatic, but this matters.

Their fascination leans towards flat, single-petaled flowers where they can easily access nectar. Bumblebees, equipped with their robust bodies, are drawn to tubular flowers that challenge them to burrow deep for nectar rewards. Butterflies, delicate in both form and nature, prefer clusters of tiny flowers, providing ample surface area to land and feed. Hummingbirds, with their vibrant energy, are enticed by tubular flowers rich in nectar, often painted in fiery reds and oranges, colors invisible to bees but vivid to the birds. Nobody ever told me this directly — I had to learn it.

Native Plants and Their Superpower

Native plants are the unsung heroes in the quest to attract pollinators. They are the heart song of any ecosystem, honed over millennia to thrive in local soils and climates. Their evolution alongside indigenous pollinators creates a symbiotic relationship unmatched by exotic species. It might sound dramatic, but this matters.

By incorporating native plants, you offer pollinators a habitat in full harmony with their instincts and needs. Research your local area to discover these botanical treasures. Whether it's the purple coneflower in North America, the field scabious in Europe, or the kangaroo paw in Australia, each locale boasts plants that can form the backbone of a pollinator-friendly garden. Nobody ever told me this directly — I had to learn it.

The Art of Planning Your Garden

A garden that draws in pollinators is as much about form as it is about function. Aesthetic appeal often naturally aligns with ecological value. Start by ensuring a succession of blooms throughout the growing season. It might sound dramatic, but this matters.

This ensures that from the first hint of spring warmth until the crisp touch of autumn, your garden is a continuous tapestry of color and nectar. Early bloomers, like crocuses and wild cherries, fuel the initial bust of pollinator activity after winter. As spring turns to summer, the floral parade continues with the likes of lavender and bee balm. Finally, late bloomers like asters and goldenrods offer valuable resources as insects prepare for colder months or migration. I used to completely misunderstand this.

Adjacent to your flowing beds of color, consider planting hedgerows or a mix of shrubs. Here's the truth nobody talks about.

Hawthorn, sloe, and roses provide shelter and additional food sources such as hips and haws, which support pollinators beyond the flowering season. Intersperse these with berry-producing plants like elderberry or blackthorn that feed birds but also host a suite of pollinator activity. It might sound dramatic, but this matters.

To be honest, #### The Practicalities: Plant Varieties and Maintenance

When choosing specific plant varieties, opt for open-pollinated or heirloom seeds whenever possible. Hybrid varieties, bred for aesthetics, often lack in nectar production, offering less to visiting pollinators. Let me tell you, this changed how I think.

Additionally, consider plant density and grouping. Large swaths of single species in full bloom create a visual and aromatic lure that's easy for pollinators to spot from afar. I used to completely misunderstand this.

Maintenance is another cornerstone of a pollinator-friendly garden. Avoid pesticides and chemicals that can damage sensitive pollinator populations. Let me tell you, this changed how I think.

Instead, embrace organic practices; the presence of pests can sometimes indicate a healthy ecosystem, providing food for natural predators. Mulching your garden not only conserves moisture but also nurtures beneficial soil organisms, contributing to plant health and vitality. Nobody ever told me this directly — I had to learn it.

Creating Habitats Beyond Flowering Plants

Remember, attracting pollinators involves more than just providing nectar. Birds and insects need shelter, nesting materials, and water. It might sound dramatic, but this matters.

Incorporate features like bee hotels, logs, and rock piles to furnish these necessities. A water source, such as a shallow bird bath or a small pond, serves as a hydration station and a draw for dragonflies, indispensable guardians against pests like mosquitoes. It might sound dramatic, but this matters.

Celebrating Your Efforts and Observations

Many people don't realize this, but Your work, though rooted in the soil, blooms in inspiration. As your pollinator sanctuary flourishes, take time to appreciate the diversity and abundance it fosters. Nobody ever told me this directly — I had to learn it.

Each visit, each pollination act, weaves a story of life and continuity. Cultivate a habit of observation—learn to identify the charming bee dances, the dynamic flight patterns of butterflies, and the mesmerizing aerobatics of hummingbirds. These daily performances will deepen your connection to the natural world and elevate your gardening efforts from pastime to passion. Nobody ever told me this directly — I had to learn it.

Ultimately, selecting the right plants for attracting pollinators transforms a simple garden into a thriving ecosystem. The choices you make ripple through your immediate environment, contributing to global sustainability. Every flower and plant meticulously chosen plays its part in an ecological symphony, celebrating the interconnectedness of all life. It might sound dramatic, but this matters.

By empowering pollinators, we secure the future of our gardens and our planet, ensuring they remain not only a home for us but a haven for all. Here lies the true beauty of gardening—a testament to growth, transformation, and the sacred dance of biological reciprocity. Embrace it wholeheartedly. Let me tell you, this changed how I think.

Chapter 7: Designing a Pollinator-Friendly Garden Layout

The gentle hum of bees, the fluttering dance of butterflies, and the sparkling glint of sunlight on the wings of hummingbirds — these are the signs that your garden is not just a static display of floral colors but a vibrant ecosystem. It might sound dramatic, but this matters.

Designing a pollinator-friendly garden layout is more than choosing the right plants; it's about crafting a space that supports and nourishes these vital creatures that play a crucial role in the health of our environment. Let me tell you, this changed how I think.

Understanding Pollinator Needs

Interestingly enough, Before laying out your garden, it's essential to get inside the feathers, wings, and busy feet of your pollinator allies. Pollinators like bees, butterflies, moths, beetles, bats, and birds each have unique requirements. It might sound dramatic, but this matters.

Creating a smorgasbord of opportunities for these creatures is the foundation for a thriving environment. They need food, water, shelter, and a safe habitat free from harmful chemicals. Here's the truth nobody talks about.

Food: Different pollinators are attracted to various flower shapes, colors, and scents. Bees favor violet, blue, and yellow flowers, while hummingbirds are drawn to red, tubular ones. I used to completely misunderstand this.

Butterflies have a penchant for vivid, bright colors like red, yellow, and orange. Ensure continuous blooms throughout the seasons so pollinators always find nectar and pollen. Here's the truth nobody talks about.

Water: Pollinators require shallow water sources. Here's the truth nobody talks about.

A simple birdbath with stones or marbles creates landing pads for these critters, or you can design a small, gentle bubbling fountain to serve as both decoration and hydration point. Nobody ever told me this directly — I had to learn it.

At first glance, **Shelter and Habitat:** Native plants are key, as they are adapted to your local environment, offering reliable habitat. It might sound dramatic, but this matters.

Providing nesting sites enhances their chances of visiting your garden. Include elements like dead wood, bare patches of earth, and piles of leaves to facilitate the nesting of solitary bees. Nobody ever told me this directly — I had to learn it.

Planning Your Space

Once you understand the basics, it's time to dream up your layout. Let me tell you, this changed how I think.

Whether you have an expansive yard or a compact urban plot, the principles remain the same. Designing for pollinators calls for strategic planning to mix aesthetic appeal with ecological function. I used to completely misunderstand this.

I've seen this happen before: **Zones and Layers:** Divide your garden into zones that mimic natural habitats. Nobody ever told me this directly — I had to learn it.

Aim for diversity in plant height, texture, and density. Taller trees and shrubs can serve as windbreakers, offering protection for more sensitive plants below and creating different layers of interest. I used to completely misunderstand this.

Clumps and Swathes: When planting, avoid isolated specimens. Nobody ever told me this directly — I had to learn it.

Group plants in community clusters or broader swathes. This provides more substantial visual targets for pollinators and encourages them to linger longer as they forage. Let me tell you, this changed how I think.

Pathways and Accessibility: Design inviting pathways that meander invitingly through your garden. I used to completely misunderstand this.

These trails shouldn't just be practical for human visitors, but also easy-to-navigate flight paths. Stick to natural materials for paths; gravel paths are excellent for drainage and offer a comfortable route for beetles and other small ground-dwelling pollinators. Here's the truth nobody talks about.

Integrating the Hardscape

Structures are not merely functional; they add character and depth to your design. It might sound dramatic, but this matters.

Incorporate trellises, arches, and pergolas, which can serve dual purposes of supporting climbing plants and providing sheltering perches for winged guests. It might sound dramatic, but this matters.

Raised Beds and Containers: In smaller spaces or areas with poor soil, raised beds or containers can help by providing optimal growing conditions. Let me tell you, this changed how I think.

They should be arranged to ensure accessibility for pollinators and complement your layout theme. Let me tell you, this changed how I think.

Let's not forget, **Water Features:** Incorporating a water feature adds a sense of tranquility and serves vital functions. Nobody ever told me this directly — I had to learn it.

Even a small feature, like a pot turned into a water garden, can attract wildlife. Keep water fresh and clean, avoiding stagnant areas where pests might breed. I used to completely misunderstand this.

Embracing the Seasons

Remember, a successful garden harmonizes with, responds to, and transforms across the seasons. Nobody ever told me this directly — I had to learn it.

Consider the timing of blooms, foliage, and seed production to maintain a dynamic and engaging landscape. I used to completely misunderstand this.

Spring Explosion: Focus on early bloomers; they provide necessary resources as pollinators emerge from dormancy. I used to completely misunderstand this.

Think of crocuses, hellebores, or snowdrops as early appetizers for your visiting pollinators. It might sound dramatic, but this matters.

Summer Abundance: This is when your garden will be in its full glory. Let me tell you, this changed how I think.

Maximize blooms through a succession of planting. Sun-loving perennials like lavender, daisies, and catmint are ideal. Here’s the truth nobody talks about.

Fall Furnishings: Late-blooming plants will be vital in extending the foraging season into the fall months. Here’s the truth nobody talks about.

Asters, goldenrod, and sedums are just a few that offer late-season sustenance. Nobody ever told me this directly — I had to learn it.

Winter Rest: Don’t overlook the colder months. Nobody ever told me this directly — I had to learn it.

Leave seed heads, such as those from coneflowers, to serve as a food source. Evergreens can provide shelter, while native grasses add textural interest. I used to completely misunderstand this.

Committing to Eco-Friendliness

Lastly, one of the greatest gifts you can offer your garden’s inhabitants is a commitment to sustainability. I used to completely misunderstand this.

Avoid chemical pesticides and fertilizers that may compromise your pollinator population. Instead, encourage a natural balance of organisms through companion planting and organic pest control methods. Here’s the truth nobody talks about.

Conclusion

Designing your pollinator-friendly garden is not just about ticking environmental boxes; it’s an opportunity to connect deeply with the rhythms of life. I used to completely misunderstand this.

As you cultivate these relationships, you’ll find yourself enchanted by the daily ballet of bees, captivated by the elegant gliding of butterflies, and soothed by the symphony of whispers in the wind, wings, and water. It might sound dramatic, but this matters.

Embark on this journey with open eyes and an open heart. Let me tell you, this changed how I think.

Over time, your garden will not only flourish in beauty and biodiversity, but it will also become a source of endless inspiration and a testament to the reciprocal bond between humanity and nature. Embrace the transformation, and in return, your garden will unlock an untamed narrative of growth, life, and harmony. It might sound dramatic, but this matters.

Chapter 8: The Art of Companion Planting

Imagine stepping into a garden where every plant is not just growing but thriving, each leaf, flower, and stem engaged in a silent symphony, an intricate dance of mutual benefit and harmonious coexistence. Nobody ever told me this directly — I had to learn it.

Such is the garden of companion planting—a technique that has been honed through centuries, drawing wisdom from nature herself. Let me tell you, this changed how I think.

Companion planting is, at its heart, the practice of growing different plants in close proximity to assist each other in various ways. Let me tell you, this changed how I think.

It's nature's way of encouraging collaboration over competition, maximizing the health and yield of your garden's bounty. Think of it as matchmaking for plants, aiming for partnerships that will weave a resilient ecosystem within your plot. Let me tell you, this changed how I think.

The appeal of companion planting goes beyond aesthetics and yields; it touches on a deeper ecological understanding. I used to completely misunderstand this.

The method embraces biodiversity, celebrates inter-dependency, and encourages sustainability. In essence, it's an acknowledgement that everything in a garden is intrinsically connected, and by nurturing these connections, we plant seeds of abundance. Let me tell you, this changed how I think.

Synergy and Support

By pairing plants that complement each other, you create an environment that minimizes the need for chemical fertilizers and pesticides. It might sound dramatic, but this matters.

For instance, the classic combination of tomatoes and basil is not just a culinary delight; it also demonstrates the power of companionship in the garden. Basil helps repel insects that typically prey on tomatoes, while the lush canopy of tomato plants offers shade, creating a microclimate that keeps the basil cool and retained with moisture. Nobody ever told me this directly — I had to learn it.

As strange as it sounds, Similarly, consider the age-old trio known as the "Three Sisters": corn, beans, and squash. Let me tell you, this changed how I think.

This Native American planting tradition showcases a perfect synergy. Corn provides a natural trellis for bean vines to climb, beans fix nitrogen into the soil, enriching it for the benefit of all three plants, and squash spreads across the ground, suppressing weeds and maintaining soil moisture. It might sound dramatic, but this matters.

Pest Control, Naturally

One of the marvelous aspects of companion planting is its ability to naturally deter pests. Here's the truth nobody talks about.

Certain plants emit scents that confuse or repel harmful insects, providing protection for their neighbors. Marigolds are celebrated champions in this arena; with their pungent aroma, they effectively deter nematodes and a plethora of other soil-borne pests, making them excellent companions for a wide range of garden vegetables. Let me tell you, this changed how I think.

Many people don't realize this, but Nasturtiums, with their vibrant blooms and peppery scent, attract aphids away from more vulnerable plants like tomatoes or roses, acting as a decoy. Here's the truth nobody talks about.

Similarly, lavender, with its delightful fragrance, not only attracts pollinators but also drives away moths, fleas, and mosquitoes, proving beneficial near crops that these pests target. Here's the truth nobody talks about.

Enhancing Growth and Flavor

Companion planting can also influence the growth rate and the taste of your vegetables. When carrots are planted with chives, the chives can deter the carrot fly, a known menace. Here's the truth nobody talks about.

Additionally, chives' presence can enhance the carrot's flavor. This flavor enhancement is not just folklore—certain chemical interactions, including the release of essential oils or nutrients, can indeed alter the taste profile of adjacent plants. Here's the truth nobody talks about.

I've seen this happen before: The nutritional improvement extends to the soil where leafier companions, such as spinach, which share a bed with beetroot, can enrich the earth as their decomposed leaves act as a green mulch, providing nutrients and improving soil structure. Here's the truth nobody talks about.

Respecting Plant Personalities

In the delicate art of companion planting, not every plant pairing is beneficial. Some plants have competitive habits that make them unsuitable neighbors. Nobody ever told me this directly — I had to learn it.

For instance, fennel and dill may seem like friendly companions, but they are best grown apart. Fennel, in particular, can inhibit the growth of many other garden plants due to allelopathic chemicals released into the soil—a stark reminder that even plants have personalities and boundaries that ought to be respected. I used to completely misunderstand this.

In real life, Similarly, tall plants with sprawling roots like sunflowers and potatoes can compete excessively for nutrients and moisture, stunting each other's growth if planted too closely. Here's the truth nobody talks about.

Creating a Pollinator's Paradise

Companion planting plays a pivotal role in attracting and supporting pollinators. I used to completely misunderstand this.

The more diverse your plant selection, the higher the likelihood of creating a thriving ecosystem of beneficial insects. Companion plants like borage and calendula not only add stunning beauty to the garden with their blue and orange blossoms respectively but also draw bees and butterflies with their rich nectar, boosting pollination rates. Here's the truth nobody talks about.

Herbs such as dill, coriander, and parsley, if allowed to flower, become magnets for predatory insects such as hoverflies, which feed on aphids, ensuring your vegetables grow undisturbed by those munching pests. Let me tell you, this changed how I think.

The Spiritual Connection

Beyond the pragmatic benefits, there is a spiritual element in companion planting that connects you more profoundly with your garden. Let me tell you, this changed how I think.

As you observe how one plant aids another, fostering welfare that extends beyond themselves, you gain a greater appreciation of the symbiotic relationships that bind all living things. It might sound dramatic, but this matters.

Interestingly enough, By viewing your garden as an ecosystem rather than isolated plants, your role shifts from mere caretaker to a steward of an intelligent, living community. Nobody ever told me this directly — I had to learn it.

This shift inspires a more profound sense of belonging and responsibility, encouraging you to listen to what your garden needs and respond with care, creativity, and respect. Let me tell you, this changed how I think.

As you embark on your journey into the world of companion planting, embrace experimentation. No two gardens are the same, and what thrives in one may falter in another. Here's the truth nobody talks about.

Pay attention to the stories your plants tell—their growth, their struggles, and their successes. With patience and curiosity, you can craft a garden that bursts with vitality, nourished by the symphony of plants living and thriving together. It might sound dramatic, but this matters.

Gardening is a continuous learning process, much like life, and companion planting invites us to learn by doing, adapting, and growing alongside our green friends. Here's the truth nobody talks about.

Each pairing you try is a new chapter in the unfolding story of your garden, enriching not just the patch of earth under your care but also your soul. Nobody ever told me this directly — I had to learn it.

In the end, companion planting is a dance, a creative partnership where soil, plants, and gardener come together, writing lines of beauty and abundance in the tapestry of earth. Nobody ever told me this directly — I had to learn it.

By nurturing these relationships, we unlock not just the potential of our gardens but also a deeper understanding of nature's nuanced elegance. Nobody ever told me this directly — I had to learn it.

Chapter 9: Creating a Year-Round Flowering Calendar

In the symphony of nature, every flowering plant plays its own unique tune. Imagine walking through your garden, enveloped in a floral symphony that changes with each passing season. Here's the truth nobody talks about.

This dream can become reality by thoughtfully curating a year-round flowering calendar that not only beautifies your garden but attracts pollinators, supports biodiversity, and enriches your gardening experience. The joyful dance of colors and fragrances throughout the year offers more than aesthetic pleasure; it fosters a thriving ecosystem that benefits both wildlife and our environment. Let me tell you, this changed how I think.

Creating a year-round flowering calendar requires a blend of art and science—understanding plant cycles, local climate conditions, and the needs of pollinators. It might sound dramatic, but this matters.

Start by embracing a diversity of plants, each with its own bloom time, to seamlessly transition from one season to the next. This ensures your garden is rarely without blossoms and continues to draw in a variety of pollinators. Nobody ever told me this directly — I had to learn it.

Spring: The Awakening

Spring marks the grand opening of the floral calendar, a renaissance after winter's dormancy. It might sound dramatic, but this matters.

It's a time when nature bursts forth in reckless abandon. Early bloomers such as crocuses, snowdrops, and daffodils are the heralds of spring, pushing their way through the remnant cold and luring the first bees of the year who forage for nectar. I used to completely misunderstand this.

As the season progresses, consider introducing tulips, bluebells, and azaleas. Their vibrant hues and bold forms capture the essence of revival. For a more exotic flair, incorporate flowering cherries and witch hazel. Let me tell you, this changed how I think.

Remember, your choices should reflect both aesthetic harmony and ecological balance. By favoring native plants, such as bloodroot and Eastern redbud, you strengthen the local ecosystem. Native species are well-adapted to local conditions and are more likely to attract native pollinators. Nobody ever told me this directly — I had to learn it.

Summer: The Full Bloom

Summer is the peak of the garden's life cycle—a magical time when plants bask in warmth and light, reaching their full splendor. Nobody ever told me this directly — I had to learn it.

This season's tapestry is woven with the bold colors of roses, the delicate forms of lilies, and the sun-drenched brilliance of sunflowers. Here's the truth nobody talks about.

Choose a variety of species with staggered bloom times for continuous flowers from June through August. Marigolds, cosmos, and lavender are great choices to keep the garden lively. Let me tell you, this changed how I think.

Including plants like echinacea and coreopsis ensures a vibrant show while supporting pollinators during this abundant period. Something worth remembering: that hydration is key for summer blooms, so implement watering strategies to sustain your plants through heat spells. It might sound dramatic, but this matters.

Autumn: The Grand Finale

As summer fades, don't let your garden's vigor wan. Let me tell you, this changed how I think.

Autumn offers a palette rich in deep reds, oranges, and golds—capturing the essence of transition and harvest. Chrysanthemums, asters, and autumn crocus brighten the garden alongside the warm tone of fallen leaves. Here's the truth nobody talks about.

Sedums and goldenrods flourish in the cooling weather, providing ample nectar for bees preparing for winter. It might sound dramatic, but this matters.

Even as the days shorten, your garden can flourish with appropriate selecions. Incorporate late-blooming salvias and Japanese anemones to extend the flowering season into late autumn, ensuring the wind-down is gradual and graceful. Here's the truth nobody talks about.

Winter: The Quiet Splendor

While many views winter as a time of dormancy, it can be a season of subtle charm if managed creatively. Let me tell you, this changed how I think.

Consider evergreens and plants with winter interest, such as hellebores, winter jasmine, and witch hazel. These hardy plants can defy the frost, presenting delicate blossoms in the stark landscape. It might sound dramatic, but this matters.

Additionally, berry bushes such as holly and beautyberry provide both aesthetic appeal and necessary food sources for birds and remaining pollinators. It might sound dramatic, but this matters.

Wrapping up cold weather preparation with mulch and protective coverings will ready your garden for newfound resilience. It might sound dramatic, but this matters.

Planning and Execution: A Gardener's Guide

The creation of a flowering calendar is as much about planning as it is about plant selection. Begin by mapping out when and where each plant will thrive best in your specific landscape. It might sound dramatic, but this matters.

Familiarize yourself with the USDA Plant Hardiness Zone Map, allowing you to choose plants that will thrive in your climate with minimal intervention. Take into account sun exposure, soil conditions, and water availability for plant grouping. Choose companion plants that can coexist mutually and support one another. It might sound dramatic, but this matters.

Also, it is vital to consider the diversity of pollinators you wish to attract. Nobody ever told me this directly — I had to learn it.

Different pollinators are active at different times, and varied plants will draw different species. For instance, bees are attracted to yellow and blue flowers, while butterflies may prefer brightly colored and heavily scented blooms with long, tubular shapes. Nobody ever told me this directly — I had to learn it.

Creating Balance and Sustainability

Sustainability should be at the core of your flowering calendar. Nobody ever told me this directly — I had to learn it.

Aim to create a garden that requires minimal artificial intervention, using organic fertilizers and natural pest management techniques. Encourage healthy soil ecosystems with composting, and save water via rain catchment and mulch to lock in moisture. It might sound dramatic, but this matters.

Creating a year-round flowering garden is a labor of love, requiring patience, curiosity, and creativity. I used to completely misunderstand this.

Every blooming petal represents a step closer to harmony with nature—your unique composition contributing to the grand opera of life just outside your door. Enjoy the journey, and witness your garden transform into a living testament to biodiversity and sustainability. Here's the truth nobody talks about.

As you weave the garden into the rhythm of the seasons, you'll find that the rewards extend far beyond visual delight. I used to completely misunderstand this.

You'll be crafting a habitat that pays homage to our planet's exquisite intricacy and beauty, a sanctuary for pollinators, and a personal refuge for the soul. It might sound dramatic, but this matters.

Chapter 10: Building and Maintaining Pollinator Habitats

Imagine stepping into your garden, where the morning sun casts a golden hue over vibrant blooms. Bees dance from flower to flower, butterflies flutter by, and birds serenade you with their morning songs. I used to completely misunderstand this.

This is not just a dream—it can be your reality. By creating and maintaining pollinator habitats, you breathe life into your garden and support the very creatures that are essential for pollination. In this chapter, we journey together to transform your garden into a thriving sanctuary for pollinators. Here's the truth nobody talks about.

Starting with the Soil: The Foundation of Life

I've seen this happen before: Healthy gardens start from the ground up. Rich, well-cultured soil is the bedrock of a flourishing pollinator habitat. Let me tell you, this changed how I think.

Aim to enrich your soil organically by incorporating compost, which not only improves soil structure but also attracts earthworms and microorganisms—small but mighty contributors to soil health. As you nurture the soil, you're establishing a robust environment where plants can thrive. Remember, healthy plants produce larger flowers that are more attractive to pollinators. Here's the truth nobody talks about.

Selecting the Right Plants: Diversity is Key

The heart of any pollinator habitat lies in the diversity of its plant life. Plant selection should be guided by the principle of diversity—offering a rich tapestry of colors, shapes, and scents. I used to completely misunderstand this.

Opt for a mix of native flowers, shrubs, and trees. Native plants are adapted to local climate conditions, require less maintenance, and provide the necessary food source for local pollinators. Nobody ever told me this directly — I had to learn it.

Believe it or not, Select flowers with different blooming periods to ensure a continuous supply of nectar and pollen throughout the growing season. Here's the truth nobody talks about.

Intentionally intersperse early bloomers like crocuses and late bloomers like asters among mid-season favorites such as coneflowers and bee balm. This ensures pollinators have a constant buffet from spring through fall. Let me tell you, this changed how I think.

Creating Shelter: Safe Harbor for Pollinators

Pollinators need more than food; they require safe spaces to rest and reproduce. Let me tell you, this changed how I think.

Introduce features that offer shelter, such as butterfly houses, bee hotels, or simply a pile of twigs and leaves for nesting. Bees, particularly solitary bees, rely on cozy nooks and crannies; consider leaving areas of bare ground for ground-nesting species. Nobody ever told me this directly — I had to learn it.

Let's not forget, Additionally, tangles of vegetation and small shrubs offer refuge from predators and extreme weather. It might sound dramatic, but this matters.

Maintaining a portion of your garden in its natural state, with its own microhabitats such as fallen leaves and logs, can further enhance these safe spaces. It might sound dramatic, but this matters.

Providing Water Sources: The Nectar of Life

Water is a crucial, often overlooked component of a pollinator habitat. Bees, butterflies, and birds all require water not just for drinking, but also for cooling their hives and dissolving minerals. Let me tell you, this changed how I think.

You can add a small, shallow birdbath or set out shallow dishes with stones for perches. To prevent drowning and provide safe access, ensure the water is no deeper than a quarter inch with plenty of landings. It might sound dramatic, but this matters.

At first glance, Consider adding a small water feature like a pond or stream, which also contributes to your garden's ecosystem by attracting yet more wildlife. Let me tell you, this changed how I think.

Such features, when placed strategically, can transform your garden into a delightful oasis for all manner of creatures. I used to completely misunderstand this.

Avoiding Pesticides: Choosing Nature

In your quest for the perfect pollinator habitat, practice restraint regarding chemicals. Many pesticides, even those labeled benign, can be harmful to pollinators. Let me tell you, this changed how I think.

Instead, embrace natural pest control techniques: introduce beneficial insects like ladybugs and lacewings, or use insecticidal soaps derived from natural sources. Trust in the resilience of nature; an ecosystem in balance often requires little interference. Here's the truth nobody talks about.

Building Connectivity: Passageways of Nectar

Pollinator habitats gain strength through connection. If space allows, create pathways or corridors connecting different planting areas with hedgerows or blooming borders. Nobody ever told me this directly — I had to learn it.

These green passageways allow pollinators safe passage across larger areas, granting them access to different nectar sources and nesting sites. These connections replicate the diversity and continuity found in nature, encouraging a biodiverse and resilient garden landscape. Nobody ever told me this directly — I had to learn it.

Embracing Imperfection: Nature's Way

Many people don't realize this, but A thriving pollinator habitat revels in the beauty of imperfection—it honors nature's organic evolution rather than chasing the notion of flawless gardens. Here's the truth nobody talks about.

Accepting wild edges, letting grass grow a bit longer, and watching where wildflowers spring up, all meld into an authentic approach to gardening. Ultimately, these aspects invite intriguing interactions and contribute to a larger tapestry of life, which in turn fosters inspiration and discovery. I used to completely misunderstand this.

Constant Curiosity and Commitment

Building a pollinator habitat is both an act of love and stewardship. Let me tell you, this changed how I think.

As you cultivate this space, observe and learn from the interactions you witness—each day yields new lessons about plant-animal dynamics, weather influences, and even the patience of a visiting monarch. Here's the truth nobody talks about.

Take notes and photographs; share your experiences and findings with like-minded gardeners and the community. Let me tell you, this changed how I think.

Engagement creates a ripple effect, inspiring others to foster pollinator habitats of their own. Nobody ever told me this directly — I had to learn it.

Towards a Resilient Future: Leaving a Legacy

Every action you take in creating a pollinator habitat contributes to a lasting legacy—a future where pollinators are free to thrive. Let me tell you, this changed how I think.

Your dedication does more than beautify a landscape; it provides vital resources needed more now than ever. With your guidance, future generations will not only inherit a world adorned with flowers and birdsong but also the wealth of knowledge you gather and share along this journey. It might sound dramatic, but this matters.

It's worth noting that The garden you nurture today casts a hopeful promise into the world—an invitation to nature to flourish, to persist, and to prosper. Here's the truth nobody talks about.

So step beyond your garden gate, embrace this quest with open arms, and let your garden grow in ways limited only by the expanse of your imagination. The world of pollinators awaits; let your garden rise to meet their call. Let me tell you, this changed how I think.

In crafting and maintaining these vibrant habitats, you're unlocking a symbiotic relationship with nature that blossoms with each season—a culmination of intentional acts rooted in respect, care, and an unwavering belief in the power of growth. Let me tell you, this changed how I think.

Here in these spaces, among the flowers and the hum of wings, humanity finds harmony with the natural world, and the garden becomes a testament to a hopeful, shared future. It might sound dramatic, but this matters.

Chapter 11: Water Features and Their Importance for Pollinators

As we turn the pages of our gardening guide to welcoming pollinators, we reach a critical, awe-inspiring element that sustains life and thrives amidst every leaf and petal—the element of water. Nobody ever told me this directly — I had to learn it.

From the gentle murmurs of a babbling brook to the serene surface of a reflective pond, water features not only enhance the aesthetic appeal of our gardens but also provide vital resources to our diligent pollinators. Let me tell you, this changed how I think.

Imagine, for a moment, stepping into your garden on a dewy morning. The sunlight dances upon droplets scattered across leaves, and as you pause, you hear the tranquil sound of water cascading over stones. Let me tell you, this changed how I think.

This enchanting aspect of your garden doesn't merely charm human senses; it serves as a beacon for bees, butterflies, birds, and an assortment of other pollinators. Let's delve into the facets of water features and their remarkable role in creating a haven both visually and ecologically vibrant. I used to completely misunderstand this.

The Magnetism of Water

Pollinators, much like all living organisms, require water to survive. Yet, the importance of water features stretches beyond their role in providing hydration. Nobody ever told me this directly — I had to learn it.

For bees, particularly honeybees, water is integral to maintaining hive humidity and for cooling purposes during sweltering summer days. When these industrious creatures approach water's surface, they do so with delicate precision, seeking shallow edges or perches that allow them to drink without the risk of drowning. Here's the truth nobody talks about.

For butterflies, water sources often serve a dual purpose. Here's the truth nobody talks about.

Many butterflies engage in a behavior known as "puddling," where they congregate on moist soil or shallow water, extracting not just water but also essential minerals to fortify their diet. Basins of gently splashing water or simple pans filled with wet mud become as attractive to them as any vibrant bloom. Nobody ever told me this directly — I had to learn it.

As strange as it sounds, Despite the allure water can offer, pollinators face peril from deep water sources with steep edges or rapid currents. Therefore, crafting water features for pollinators involves thoughtful design aimed at accessibility and safety. Here's the truth nobody talks about.

Provide shallow basins or gradually sloping edges to allow safe landings for these winged visitors. Flat stones or floating leaves offer ideal perches, keeping bees and butterflies from precarious plunges. Let me tell you, this changed how I think.

Types of Water Features and Their Benefits

Pollinator-friendly gardens can host an array of water features, each with its own charm and benefits. It might sound dramatic, but this matters.

From simple, budget-friendly additions to intricate, artful installations, the options are manifold. I used to completely misunderstand this.

I've seen this happen before: **Birdbaths:** One of the most accessible water features, birdbaths serve dual purposes. Here's the truth nobody talks about.

Not only do they provide essential hydration and bathing spots for our feathered friends, but strategically placed pebbles within their basins offer ideal drinking stations for bees and butterflies. Ensure you refresh the water daily, particularly in warm seasons, to maintain cleanliness and attractiveness. Nobody ever told me this directly — I had to learn it.

Small Ponds: For those with space, a small pond can transform the landscape into a thriving ecosystem. Ponds, with their diverse plantings and habitats, draw a wide range of pollinators and wildlife. Let me tell you, this changed how I think.

Marginal plants can provide natural landings and are critical in controlling flooding by absorbing excess water. However, ensure these oases contain shallow portions and are fish-free, as fish can predate upon larvae and pupae of important pollinators. It might sound dramatic, but this matters.

Dripping or Bubbling Rocks: These installations offer both dynamic aesthetic appeal and security for pollinators. Here's the truth nobody talks about.

The gentle drip or bubble of water over rocks makes hydration accessible without requiring the creatures to risk venturing into deeper waters. Naturally, their movement also acts as a sound attractant to pollinators in search of this vital resource. It might sound dramatic, but this matters.

Let's not forget, **Rain Gardens:** With growing awareness around water conservation, rain gardens have gained popularity. I used to completely misunderstand this.

These gardens, designed to collect runoff from roofs and driveways, hold and gradually absorb the water, reducing stormwater impact while supporting pollinator habitat. The diverse plantings in rain gardens often host pollinators themselves, further intertwining sustenance and aesthetics. Here's the truth nobody talks about.

Creating a Sanctuary: Environmental Considerations

When designing water features, it's important to consider the broader environmental impacts and responsibilities. Nobody ever told me this directly — I had to learn it.

Use non-toxic materials and avoid treatments, such as chemical algaecides, which can deter and harm pollinators. Encourage diverse plantings near the water's edge to mimic natural environments, offering additional nectar and shelter options. I used to completely misunderstand this.

Interestingly enough, Furthermore, water features provide opportunities for education and connection within a community. I used to completely misunderstand this.

Integrating interpretative signage detailing the role of pollinators and water conservation can transform a garden visit into an enlightening experience, fostering stewardship toward these vital creatures. Let me tell you, this changed how I think.

Finally, remember that water features, however small, connect your garden to larger ecosystems. It might sound dramatic, but this matters.

They may become stepping stones for migrating pollinators, linking them from one habitat to the next. By thoughtfully incorporating these features, you contribute to a collective endeavor of ecological restoration and protection. Here's the truth nobody talks about.

A Source of Renewal

Gardens built for pollinators are spaces of harmony where the delicate interaction between flora, fauna, and elements creates a self-sustaining cycle of life. I used to completely misunderstand this.

Water features are the lifeblood of these spaces, ensuring that the hearts and wings that traverse them continue to beat strong. As you nurture these features, you cultivate a sanctuary not just for pollinators, but also for your own spirit, drawn to the serene beauty and rhythmic song of water. Here's the truth nobody talks about.

As stewards of the land, let us embrace our role and cherish every ripple, every droplet needed by the vibrant wings that grace our skies. Let me tell you, this changed how I think.

Through water features, may we unlock a garden that blooms with life and invites all who wander by to linger in its embrace—a testament to the wondrous connections we cultivate when we open our landscapes to the thriving spectacle of nature's wonders. I used to completely misunderstand this.

Chapter 12: Natural Pest Control: Encouraging Beneficial Insects

Imagine stepping into a vibrant garden awash with color, buzzing with the hum of life, and thriving under the attentive care of nature itself. I used to completely misunderstand this.

Here, the unseen hands of industrious insects maintain a delicate balance, one that keeps destructive pests at bay and ensures the flourishing of plant life. In this chapter, we explore the fascinating world of beneficial insects and how you can encourage them to take up residence in your own garden, creating a self-regulating ecosystem that reduces the need for chemical interventions. Let me tell you, this changed how I think.

The Secret Guardians of the Garden

Interestingly enough, While some insects may seem like the bane of a gardener's existence, others are secretly working for you, playing crucial roles in controlling pest populations and promoting plant health. Let me tell you, this changed how I think.

These beneficial insects, such as ladybugs, lacewings, and parasitic wasps, are nature's pest control agents. They prey on harmful bugs that can wreak havoc on your plants, maintaining a natural balance that supports a thriving garden ecosystem. Nobody ever told me this directly — I had to learn it.

Ladybugs, with their distinctive red and black spotted wings, are voracious predators of aphids and other soft-bodied insects. I used to completely misunderstand this.

Lacewing larvae, often referred to as "aphid lions," are equally beneficial, consuming vast numbers of aphids, thrips, and caterpillars. Meanwhile, tiny parasitic wasps lay their eggs inside or on the larvae of pests like caterpillars and whiteflies, eventually eliminating them from the inside out. Let me tell you, this changed how I think.

Creating a Hospitable Habitat

As strange as it sounds, To attract these invaluable allies, your garden needs to offer a welcoming environment that meets their basic needs: food, water, and shelter. Here's the truth nobody talks about.

Consider planting a diverse range of flowering plants that bloom at different times throughout the growing season. Diversity ensures that beneficial insects have a consistent supply of nectar and pollen, which many adults require for sustenance before and after their predatory activities. I used to completely misunderstand this.

Incorporate plants such as dill, fennel, and coriander, which are known to attract lacewings and parasitic wasps. Nobody ever told me this directly — I had to learn it.

Sunflowers and marigolds, besides adding beauty to your garden, will also attract ladybugs. Another simple way to support your insect allies is by allowing some of your herbs and vegetables to bolt and flower, providing additional nectar sources. Let me tell you, this changed how I think.

Beyond food, insects need shelter to protect them from weather extremes and predators. I used to completely misunderstand this.

Provide a range of habitats, such as rock piles, mulch layers, and even small, insect-friendly structures known as bug hotels, to give these beneficial creatures a safe place to thrive. These structures can be easily made from recycled materials—stack some wooden pallets, fill them with straw and bamboo canes, and watch as the residents move in. Nobody ever told me this directly — I had to learn it.

Reduce Chemical Use

Creating a garden that supports beneficial insects requires a commitment to minimizing or eliminating the use of synthetic pesticides and herbicides. Nobody ever told me this directly — I had to learn it.

These chemicals don't discriminate, often killing helpful insect species alongside their pest targets. Instead, opt for integrated pest management practices, which focus on long-term prevention and control through a combination of physical, cultural, and biological strategies. Nobody ever told me this directly — I had to learn it.

Remember, the presence of some pests in your garden is vital, as they provide nourishment for predatory insects. I used to completely misunderstand this.

By allowing a small number of pests to remain, you'll ensure that beneficial insects have a consistent food source, compelling them to stay and patrol your garden diligently. I used to completely misunderstand this.

Understanding the Balance

One of the most remarkable aspects of encouraging beneficial insects is watching nature strike its balance. The relationship between pests and their predators is a dynamic one; populations will fluctuate seasonally, but over time, your garden will become a self-sustaining ecosystem. I used to completely misunderstand this.

Have faith in the process and understand that some patience may be required. It is a testament to nature's wisdom—a gentle reminder that sometimes intervention is best left to the experts, those in tiny armored shells and delicate wings. I used to completely misunderstand this.

Observing and Learning

As strange as it sounds, Spend time observing the intricate world at play within your garden. Discovering which insects are gracing your green space isn't just informative; it's enchanting. I used to completely misunderstand this.

Identifying insects can be done with field guides or apps on your smartphone. Understanding which species are present and what they're doing can provide insights into how to tweak your garden to better accommodate them. Nobody ever told me this directly — I had to learn it.

Engaging children or friends in these discoveries provides shared excitement as you witness how simple changes transform the garden into a bustling hub of biodiversity. I used to completely misunderstand this.

Suddenly, spotting a ladybug is no longer mundane—it's a hopeful sign that your eco-friendly efforts are paying off. Here's the truth nobody talks about.

The Broader Impact

Let's not forget, Promoting beneficial insects in your garden resonates beyond your own space. By avoiding harsh chemicals and fostering biodiversity, you contribute to the health of the wider ecosystem. Let me tell you, this changed how I think.

Beneficial insects don't recognize property lines; they travel, seeking fertile grounds and spreading their influence. Therefore, your garden becomes an integral part in the patchwork of local habitats, supporting pollinators, birds, and other wildlife. I used to completely misunderstand this.

Moreover, nurturing beneficial insects is an effective step toward sustainable gardening practices that respect and mirror natural processes. It might sound dramatic, but this matters.

As our world becomes increasingly urbanized, gardens can be sanctuaries for these mini ecosystems. We become custodians of these spaces, playing a small but essential role in preserving the delicate balance upon which life depends. I used to completely misunderstand this.

A Call to Action

As you nurture your garden with the care and intention it deserves, remember each element is interconnected. Nobody ever told me this directly — I had to learn it.

With your guidance—providing diverse plantings, ensuring natural habitats, and reducing chemical usage—your garden becomes not only a haven for beneficial insects but also a personal sanctuary where you find peace and inspiration. Nobody ever told me this directly — I had to learn it.

Embrace the transformative power of nature's little helpers. I used to completely misunderstand this.

Encourage them, and they will reward you with a thriving garden, bursting with life and beauty. Your garden will stand as a testament to what is possible when we trust the cycles of nature—the ultimate gardener—and celebrate the wisdom of life's small creatures in their ceaseless dance towards balance and abundance. Here's the truth nobody talks about.

Chapter 13: Organic Gardening Practices for a Healthy Garden

Imagine stepping into your garden and feeling a profound sense of balance and harmony. The blossoms and leaves rustle with the gentle help of a breeze, while bees diligently buzz as they hop from flower to flower. Here's the truth nobody talks about.

Every inch of soil is alive with organisms working beneath the surface, unseen yet impactful. This is the magic of an organic garden—a haven that thrives without synthetic intrusions, resonating with the rhythms of nature. I used to completely misunderstand this.

Organic gardening is more than a method; it's a philosophy, a commitment to nurturing life in its purest form. I used to completely misunderstand this.

By embracing organic practices, we not only cultivate the land, but we also cultivate ourselves, fostering mindfulness, patience, and respect for nature's intricate systems. This chapter will guide you through the key steps to establish and maintain a healthy organic garden, unleashing an oasis that attracts pollinators and promotes robust growth. It might sound dramatic, but this matters.

Building Healthy Soil: The Lifeline of Your Garden

The cornerstone of any flourishing garden is its soil. Organic gardening emphasizes enhancing soil health without the use of chemical fertilizers. Begin by testing your soil to understand its pH and nutrient profile. Let me tell you, this changed how I think.

With this knowledge, enrich your soil with organic matter. Compost is your best ally, composed of decomposed plant material that enriches the soil with essential nutrients. Aged manure and organic mulch are also invaluable additions that improve soil texture, retain moisture, and suppress weeds. I used to completely misunderstand this.

The secret to thriving plants lies in fostering a vibrant community of microorganisms and earthworms. Here's the truth nobody talks about.

They aerate the soil naturally, improve drainage, and transform organic matter into plant-available nutrients. Encourage these life forces by minimizing soil disturbance and avoiding synthetic chemicals that can harm them. It might sound dramatic, but this matters.

Embrace Natural Pest Control

Pests are inevitable in any garden, but they can be managed effectively through organic means. I used to completely misunderstand this.

The first step is to cultivate biodiversity. An array of plant species confuses and deters pests, while inviting beneficial insects such as ladybugs, lacewings, and predatory beetles that naturally control pest populations. It might sound dramatic, but this matters.

Introduce companion planting to your garden strategy. Marigolds, for instance, repel nematodes and attract pollinators, while basil deters flies and mosquitoes. I used to completely misunderstand this.

Regular inspection of your plants helps you detect early signs of pest infestations before they escalate. Rather than reaching for synthetic pesticides, consider organic alternatives like neem oil, diatomaceous earth, or insecticidal soap. Let me tell you, this changed how I think.

Cultivate Resilient Plants

Strong plants are the backbone of an organic garden. Start by selecting native plant varieties well-suited to your climate and soil conditions. Nobody ever told me this directly — I had to learn it.

These plants are inherently more resistant to local pests and environmental stresses. Rotate crops yearly to disrupt pest and disease cycles and to maintain soil fertility. This practice reduces dependency on fertilizers and enhances biodiversity in your garden. Let me tell you, this changed how I think.

Water your plants wisely. Overwatering not only wastes resources but can also weaken plants and promote diseases. Let me tell you, this changed how I think.

Use drip irrigation systems or soaker hoses to deliver water directly to the root zone, minimizing evaporation. Mulching further aids water retention and regulates soil temperature. It might sound dramatic, but this matters.

Encourage Pollinators: Garden Allies

Pollinators are vital allies in the organic garden, responsible for the bounty of fruits, vegetables, and flowers. Let me tell you, this changed how I think.

Create a welcoming environment by planting a diverse range of flowering plants that bloom across the seasons. Different colors, shapes, and nectar provide sustenance to various pollinator species, from bees to butterflies and hummingbirds. It might sound dramatic, but this matters.

Avoid using chemical sprays that can harm these beneficial creatures. Nobody ever told me this directly — I had to learn it.

Instead, provide water sources, like shallow dishes with stones, for pollinators to drink from, and incorporate plants like milkweed and coneflower that double as food and habitat. I used to completely misunderstand this.

Composting: Nature's Recycling

Composting is a cornerstone of organic gardening, turning kitchen scraps and yard waste into "black gold. " Not only does this practice reduce waste, but it also results in a nutrient-rich amendment for your garden. Here's the truth nobody talks about.

Start with a balance of green materials (vegetable peels, coffee grounds) and brown materials (leaves, straw). Turn the compost regularly to enhance aeration and decomposition. Here's the truth nobody talks about.

A well-maintained compost pile generates heat, decomposing matter efficiently and killing pathogens. I used to completely misunderstand this.

Within weeks, you'll have a rich, earthy humus that can be spread in your garden beds, nurturing plant growth and diet-enriching your garden's soil biome. It might sound dramatic, but this matters.

Maintain Garden Health Through Observation

An organic gardener is an observer, ever attuned to the subtle cues of their green space. Regular walks through your garden offer insights into your plants' health and growing conditions. Let me tell you, this changed how I think.

Look for signs of nutrient deficiencies, pests, or disease. Changes in leaf color, wilting, or unusual spotting may indicate a need for attention. Nobody ever told me this directly — I had to learn it.

Utilize natural remedies whenever possible. I used to completely misunderstand this.

For example, yellowing leaves might require an infusion of compost tea or a sprinkle of wood ash to restore balance. Fungal infections can often be halted with a baking soda spray or by adjusting watering practices to reduce moisture on foliage. Nobody ever told me this directly — I had to learn it.

Let me leave you with this thought:, organic gardening is a journey of collaboration with nature. It calls for intention, observation, and a willingness to support life in its diverse forms. Nobody ever told me this directly — I had to learn it.

By choosing organic practices, gardeners become stewards of the earth, preserving its vitality while enjoying a lush, productive garden that teems with life. As you unlock the potential of your organic garden, you contribute to a sustainable future, one plant, and one pollinator at a time. Let this chapter inspire you to create your own green sanctuary, a testament to the harmony between humanity and the natural world. It might sound dramatic, but this matters.

Chapter 14: Pollinator-Friendly Gardening Tools and Techniques

Gardening is a symphony of harmony, a beautiful concert where every plant, insect, and drop of water plays a critical role. The garden, when carefully tended, becomes not just a place of serene beauty but a thriving habitat that invites the essential dance of pollinators. Nobody ever told me this directly — I had to learn it.

These creatures, from bees and butterflies to hummingbirds and beetles, are the unsung heroes of our botanical sanctuaries, ensuring the continuity and bounty of our labors. But how do we craft a garden that rolls out the green carpet for these winged visitors? Let us delve into the arsenal of pollinator-friendly gardening tools and techniques that transform our backyards into flourishing ecosystems. Here's the truth nobody talks about.

The Essential Tools for a Pollinator Paradise

I've seen this happen before: The heart of a pollinator-friendly garden lies in thoughtful planning and execution. Here's the truth nobody talks about.

Tools are our allies, helping us cultivate an environment where pollinators can thrive. As you embark on this journey, invest in tools that respect both the land and its little inhabitants. I used to completely misunderstand this.

1. Hand Trowels and Hoes:

Believe it or not, Opt for hand tools over large mechanical ones. Here's the truth nobody talks about.

Hand trowels and hoes allow for more precise gardening without disrupting the delicate habitats that pollinators might use for nesting or resting. Look for ergonomic designs that reduce strain, ensuring that long hours spent in the garden are comfortable and enjoyable. I used to completely misunderstand this.

2. Lightweight Hose Systems:

In real life, Water is life, and a garden must be well-irrigated to attract pollinators. Choose hoses that are lightweight and equipped with adjustable nozzles for controlled watering. Here's the truth nobody talks about.

A gentle, rain-like setting can prevent harming delicate flowers and the creatures that visit them. Consider soaker hoses or drip irrigation systems as these sustainable options limit water wastage and nurture the soil over time. I used to completely misunderstand this.

3. Compost Bins:

Many people don't realize this, but Healthy soil teems with life and offers an open invitation to pollinators. Creating a compost system is a critical step in maintaining soil vitality. Nobody ever told me this directly — I had to learn it.

Select a compost bin that suits your space—tumbling models for frequent aeration, or open-frame designs for larger volumes. Compost enriches the soil, inherently supporting a biodiverse garden landscape and enhancing nectar-rich plant growth. Nobody ever told me this directly — I had to learn it.

4. Bee Hotels and Butterfly Houses:

Pollinator homes are equally important as the blooms themselves. Invest in bee hotels and butterfly houses made from sustainable materials. Let me tell you, this changed how I think.

Position these shelters in sunny spots protected from wind to provide safe havens where pollinators can rest and reproduce. Making your garden a refuge encourages these vital populations to settle and interact regularly within your garden. It might sound dramatic, but this matters.

5. Natural Mulches:

I've seen this happen before: Mulching not only conserves moisture and suppresses weeds but also fosters a microhabitat for pollinators and the organisms they depend on. I used to completely misunderstand this.

Choose organic mulches such as straw, shredded leaves, and wood chips, which break down naturally to continue nourishing the soil. Here's the truth nobody talks about.

Techniques to Transform Your Garden

Complementing the right set of tools with effective techniques can significantly enhance the allure of your garden for pollinators. Here are some practices to perfect:

1. Plant Diversity:

To attract a range of pollinators, your garden must be a mosaic of diverse plant species. Incorporate a variety of flowering plants that bloom at different times of the year. Let me tell you, this changed how I think.

Include native species, as these are better adapted to the local climate and more enticing to native pollinators. Remember, a combination of shapes, colors, and heights not only appeals visually but meets the foraging needs of different pollinator species. It might sound dramatic, but this matters.

2. Maintain Bloom Continuity:

Strategically plan your plantings to ensure that something is always in bloom. Nobody ever told me this directly — I had to learn it.

This continuous food supply keeps your garden bustling with activity throughout the seasons. Early-blooming bulbs, mid-summer perennials, and fall flowers work in concert to attract continuous visits from eager pollinators. Nobody ever told me this directly — I had to learn it.

3. Avoid Pesticides:

Chemical pesticides are lethal to pollinators. Nobody ever told me this directly — I had to learn it.

Shift your focus toward organic pest control methods, such as encouraging beneficial predators like ladybugs and using neem oil or insecticidal soap when necessary. This way, you protect your plants from pests without risking harm to pollinators. It might sound dramatic, but this matters.

4. Provide Water Sources:

Pollinators need water just as much as plants do. Nobody ever told me this directly — I had to learn it.

Install a bird bath, shallow dishes, or small ponds with rocks and twigs for safe perching. Keep these water sources clean and filled, offering your buzzing visitors a place to hydrate. I used to completely misunderstand this.

5. Implement No-Dig Gardening:

This technique minimizes soil disruption, preserving the critters and fungi that play a role in the pollination process. I used to completely misunderstand this.

It involves adding organic layers on top of the soil, creating a natural, rich substrate that enhances the garden's resilience and productivity. Nobody ever told me this directly — I had to learn it.

The Art of Observation and Adaptation

Perhaps the most important technique of all is one that requires no particular tool but sharpens with practice: observation. Nobody ever told me this directly — I had to learn it.

Spend time in your garden, keenly watching which plants the pollinators prefer and how they interact with their environment. This knowledge is invaluable, allowing you to adapt your gardening practices to better support these essential creatures. Here's the truth nobody talks about.

As you nurture this vibrant habitat, notice the increased numbers and species of pollinators that visit your oasis. Nobody ever told me this directly — I had to learn it.

Remember, a pollinator-friendly garden is an evolving masterpiece, a testament to nature's resilient beauty harmonized with human custodianship. Nobody ever told me this directly — I had to learn it.

I've seen this happen before: In the end, pollinator-friendly gardening is not just about growing plants—it's about fostering life and cultivating hope. Each flower that blooms and each bee that buzzes reflects the endless possibilities within our reach. I used to completely misunderstand this.

Through conscious gardening, we contribute to the health of our ecosystems, ensuring future descendants inherit a world rich in diversity and teeming with life. Armed with the right tools and techniques, your garden will become a beacon for pollinators and a source of unending joy and inspiration. I used to completely misunderstand this.

Chapter 15: Community and Urban Gardens: Expanding Pollinator Impact

As the sun peeks over the genteel horizon, the city awakens in a symphony of activity. The streets hum with the rhythm of life, an orchestra of engines and footsteps. I used to completely misunderstand this.

Though urban areas seem like an unlikely sanctuary for nature, they hold seeds of potential for nurturing vibrant ecosystems within the hustle and bustle—a prospect lying in the hands of community and urban gardens. These green oases can become pivotal in expanding the impact of pollinators, bridging the divide between concrete jungles and flourishing biodiversity. Nobody ever told me this directly — I had to learn it.

Urban gardens are more than just visually appealing spaces; they are pockets of resilience and sustainability. Let me tell you, this changed how I think.

Each raised bed and blooming pot act as microhabitats, providing refuge and resources for pollinators. In sprawling cities, where traditional gardens are sparse, these environments offer vital pitstops for buzzing bees, flitting butterflies, and other hardworking pollinators as they make their way through an otherwise barren terrain. Nobody ever told me this directly — I had to learn it.

To be honest, Creating a garden in a densely populated area requires intention and vision. Start by selecting plants that serve dual roles—those that are both aesthetically pleasing and beneficial to pollinators. I used to completely misunderstand this.

Opt for a mix of native flowers, herbs, and vegetables that can thrive in various conditions, from rooftop gardens to vacant lots transformed into verdant patches. Native plants are particularly advantageous, as they've co-evolved with local pollinators and provide the most reliable sources of nectar and pollen. I used to completely misunderstand this.

One powerful example of plant synergy can be found in the strategic use of herbs. Lavender, basil, and mint not only flavor our dishes but also attract beneficial insects with their intoxicating scents. Nobody ever told me this directly — I had to learn it.

Consider interspersing these aromatic allies among more prominent flowers like sunflowers, echinacea, and cosmos. This blend creates a dynamic tapestry of colors and fragrances that beckons pollinators from miles around, creating a thriving metropolis of biodiversity. Let me tell you, this changed how I think.

Community and urban gardens also offer profound educational opportunities. They can be integrated into school programs, teaching children about ecosystems, sustainability, and the life cycles of both plants and insects. Here's the truth nobody talks about.

These gardens act as interactive classrooms, where lessons extend beyond textbooks to tangible activities such as planting, tending, and observing. Engaging the younger generation instills an appreciation and understanding that can ripple outwards, fostering a community interested in preserving natural habitats. Here's the truth nobody talks about.

Many people don't realize this, but Collaborative elements extend further than education. Community gardens thrive on the principles of cooperation and shared stewardship. Here's the truth nobody talks about.

They bring together diverse groups of people who might not traditionally engage in ecological initiatives, but who can unite over a shared interest in gardening. This collective care fosters a spirit of community, transcending individual efforts. As more areas are transformed into pollinator havens, the network of support grows, encouraging wider ecological benefits. Here's the truth nobody talks about.

To maximize these efforts, consider designing gardens with layered blooms that provide continuous blooms from early spring to late fall. This ensures a consistent food source for pollinators throughout the growing season. Nobody ever told me this directly — I had to learn it.

Additionally, leave patches of bare ground for ground-nesting bees and consider erecting bee hotels made from simple materials for solitary bee species. Water features, though they need not be large, can offer crucial hydration, while strategic planting can provide shaded areas that offer relief during the heat of the day. I used to completely misunderstand this.

As we cultivate these intentional environments, it's crucial to consider human impact and avoid using pesticides, which can inadvertently harm the very creatures we're trying to attract. I used to completely misunderstand this.

Opt for natural pest control methods, such as introducing ladybugs or using insect-repelling plants like marigolds. Here's the truth nobody talks about.

Urban gardens do more than support pollinators—they also bolster the well-being of the local human community. These green sanctuaries can improve air quality, provide fresh produce, and act as vital community gathering places. I used to completely misunderstand this.

The simple act of getting one's hands in the soil can have profound impacts on mental health, reducing stress and encouraging mindfulness. For those without personal outdoor spaces, these communal plots become crucial for accessing the therapeutic benefits of nature. Nobody ever told me this directly — I had to learn it.

Moreover, urban gardens inspire a shift in perception. Here's the truth nobody talks about.

They challenge the notion of cities as devoid of nature and instead reimagine them as hubs of biodiversity. Every time a bee lands on a city blossom, or a butterfly flutters through an unexpected greenway, it acts as a reminder of our capacity to harmonize with the natural world, even within the confines of a densely built environment. Let me tell you, this changed how I think.

Finally, as we endeavor to expand pollinator impact through urban gardening initiatives, it's essential to recognize the importance of advocacy and policy. Nobody ever told me this directly — I had to learn it.

Support can range from petitioning for more green spaces, promoting rooftop gardens, to advocating for sustainable city planning that integrates natural habitats into urban landscapes. When communities unite around these environmental goals, they can influence decision-making on a broader scale. It might sound dramatic, but this matters.

Many people don't realize this, but Let me leave you with this thought:, community and urban gardens hold immense potential to catalyze change. They are reminders that even in the heart of urbanity, we can nurture areas where life's delicate dance flourishes. It might sound dramatic, but this matters.

By coming together to cultivate these spaces, we not only expand the impact of pollinators but also send a powerful message of stewardship and hope to future generations. As these gardens grow, so too does our impact, weaving threads of green into the urban tapestry for all living beings to cherish and enjoy. I used to completely misunderstand this.

Chapter 16: The Science Behind Pollination and Plant Growth

In the intricate dance of nature, every movement has a purpose, a story, and a master plan. We often marvel at the sight of a bee flitting from bloom to bloom, a butterfly gently landing on a daisy, or the vibrant colors and enchanting fragrances that lure these pollinators to our gardens. It might sound dramatic, but this matters.

But beneath this seemingly whimsical performance lies a vast symphony of biological processes that transform a simple act into the cornerstone of our planet's flourishing ecosystems. Understanding the science behind pollination and plant growth is akin to discovering the blueprint of life itself—a blueprint that we, as gardeners and stewards of nature, are privileged to nurture and protect. Let me tell you, this changed how I think.

Pollination is the quintessential act of reproduction in the plant kingdom. At its core, pollination involves the transfer of pollen grains, the male gametes, from the anther (the male part of a flower) to the stigma (the female part). Here's the truth nobody talks about.

This journey, which may seem straightforward, is a marvel of evolutionary innovation. Each step is meticulously designed to ensure that plant species not only survive but thrive across generations. Let me tell you, this changed how I think.

Pollen grains, although microscopic, are powerhouses of genetic material. Each grain carries the potential for new life, a promise of continuity for the plant species. It might sound dramatic, but this matters.

These grains are uniquely designed to adhere to the bodies of pollinators—birds, bees, butterflies, moths, bats, and even the wind—depending on the specific mechanism of pollination a plant has evolved to employ. The diverse array of strategies plants use to attract pollinators, from the bee-friendly ultraviolet patterns on flowers to the strong, sweet scents that draw moths under the moonlight, is testament to the ingenuity of nature. Nobody ever told me this directly — I had to learn it.

The act of pollination itself is only one half of the story. Once a pollen grain successfully lands on a compatible stigma, it germinates, and a pollen tube begins to grow. Nobody ever told me this directly — I had to learn it.

This is where the journey becomes a story of resilience and precision. The pollen tube pushes its way down the style, a narrow column leading to the ovary, where the female gametes await. This journey can be fraught with obstacles, but every inch traveled is a triumph of nature's engineering. Let me tell you, this changed how I think.

Here, in the ovary, lies the crux of life's continuity: the fusion of male and female gametes, resulting in the formation of seeds. Nobody ever told me this directly — I had to learn it.

Each seed is an embryo, wrapped in a protective coat, containing the nutrients it needs to start its own life. The science of seed development is a leap of faith made possible by the marriage of genetic material and parental investment. Here's the truth nobody talks about.

It's worth noting that Plant growth beyond pollination is a saga of sunlit aspirations and earth-bound roots. With the seed as the protagonist, each plant narrates its unique story of survival, adaptation, and growth. It might sound dramatic, but this matters.

Germination is the first chapter—a dynamic rebound between environmental conditions and genetic programming. When the conditions are right—a balanced temperature, sufficient moisture, and enough light—a seed's dormant state awakens, and roots anchor into the soil while shoots reach for the sky. Here's the truth nobody talks about.

The plant world operates on delicate balances and complex feedback systems. Photosynthesis, the lifeline of plant growth, is a miracle in its simplicity and effectiveness. Let me tell you, this changed how I think.

Chlorophyll, the green pigment that gives leaves their distinctive hue, absorbs sunlight—a process that marks the transformation of light energy into chemical energy. This energy metamorphoses water and carbon dioxide into glucose, the primary food source for plants, fueling their growth, tissue repair, and development. I used to completely misunderstand this.

Yet, as any gardener knows, plants do not exist in isolation. Their growth is heavily influenced by the environment in which they dwell. Here's the truth nobody talks about.

Soil quality, nutrient availability, water supply, and climatic conditions shape the health and vitality of plants. The secret lies in understanding these factors and their interplay, respecting them, and working in harmony with them. By nurturing the roots, testing soil acidity, ensuring adequate water drainage, and responding appropriately to signs of nutrient deficiencies, gardeners can optimize the conditions required for robust plant growth. It might sound dramatic, but this matters.

Moreover, nature's narrative is enriched by the concept of mutual aid—a testament to the interconnectedness of life. Mycorrhizae, the subterranean networks that bind plant roots with fungal filaments, are a living symbiosis that enhances plant nutrient absorption and boosts soil health. Here's the truth nobody talks about.

This intricate fusion invites gardeners to explore the possibilities of fostering natural environments where these relationships can thrive. To embrace pollination is to invest not only in the beauty of the blooms today but also in the promise of future harvests, both of bounty and biodiversity. Let me tell you, this changed how I think.

In your garden, each act of facilitating pollination and encouraging plant growth is a chapter you write in collaboration with nature. Nobody ever told me this directly — I had to learn it.

By understanding the science behind these processes, you become part of a timeless story—one of life, vitality, and the relentless pursuit of flourishing. As you plant each seed, as you witness each bloom, remember that your role is that of a maestro, orchestrating a symphony of growth, beauty, and harmonious existence. Here's the truth nobody talks about.

Let your garden be a testament to the power of knowledge intertwined with the creativity of cultivation—a vibrant example of what happens when humanity's curious nature unites with the wisdom of the natural world. It might sound dramatic, but this matters.

□ Acknowledge the small actors—the bees, the wind, the droplets of rain—and realize that it is precisely these seemingly small contributors that create the grandest narratives. Here's the truth nobody talks about.

Interestingly enough, In every petal, in every leaf, in every pollinator's flight, the garden reveals a universe of wonders. Let me tell you, this changed how I think.

To unlock it is to partake in the everlasting dance of life, to plant the seeds of tomorrow on the foundations of understanding today. With awe and reverence, nurture what grows, and rejoice in the countless stories that together weave the magic tapestry of our vibrant, living world. Here's the truth nobody talks about.

Chapter 17: Monitoring and Documenting Garden Biodiversity

There exists a certain magic in gardens, woven through the buzzing of bees, the flutter of butterflies, and the rustle of life beneath the leaves. Nobody ever told me this directly — I had to learn it.

These spaces of tranquility not only offer solace to the human spirit but serve as bustling microcosms of biodiversity. To truly unlock the endless potential of your garden, it is vital to embrace the art and science of monitoring and documenting its biodiversity—a venture that transforms the act of gardening into an enlightening journey. Nobody ever told me this directly — I had to learn it.

Why, you may ask, should one delve into the meticulous task of observing garden biodiversity? Therein lies the joy of discovery and the power of awareness. Let me tell you, this changed how I think.

Monitoring allows the gardener to witness firsthand the myriad interactions between flora and fauna. It unveils the dynamics of ecosystem health and resilience, providing insights that guide conscious decision-making and foster a deeper connection to nature. Let me tell you, this changed how I think.

Let's not forget, Embarking on this endeavor begins by cultivating the mindset of an observer. Every leaf, petal, and insect carries stories, each ripe for discovery. Equip yourself with a simple journal or a dedicated app to begin documenting your observations. I used to completely misunderstand this.

Start by noting the different species of plants you have curated. These will form the foundation of your garden's ecosystem. List their flowering times, structures, and any unique characteristics that catch your eye. Let me tell you, this changed how I think.

Next, turn your attention to the faunal visitors who grace your garden—a parade of pollinators, predators, and decomposers. Establish a regular routine of observation, ideally across different times of the day and varying seasons, to capture the ebb and flow of wildlife activity. I used to completely misunderstand this.

Start with the winged wonders: bees, butterflies, and birds. Pay attention to the flowers they prefer and the times they visit. Note their behaviors—are they solitary or social, frenzied or graceful?. It might sound dramatic, but this matters.

In observing insects, recognize the less conspicuous yet equally vital creatures like beetles, ants, and spiders. Let me tell you, this changed how I think.

Each plays a crucial role in your garden's ecological pyramid. Listen to the symphony of sounds at dusk as crickets chirp and frogs emerge, an auditory testament to the hidden life within your green expanse. Nobody ever told me this directly — I had to learn it.

Many people don't realize this, but Photographic documentation serves as a powerful ally in this venture. Armed with a camera or even a smartphone, capture the dance of a bee as it gathers pollen, the vibrant flutter of butterfly wings, or the intricate web spun by a garden spider. Nobody ever told me this directly — I had to learn it.

These images become visual narratives, enriching your records and allowing for comparisons over time. Through these snapshots, patterns begin to emerge: the seasonal cycles of blooms, migration windows, and the symbiotic relationships forming the basis of your garden's natural harmony. I used to completely misunderstand this.

Beyond the qualitative aspects, consider harnessing quantitative methods to deepen your understanding. Let me tell you, this changed how I think.

Simple metrics such as counting the number of species, individuals, or particular interactions during your observation periods can provide insights into the richness and complexity of your garden ecosystem. Try setting up a small experimental plot where you can vary certain conditions—adding a new pollinator-friendly plant, for example—and monitor the effects on visitation rates. It might sound dramatic, but this matters.

As your dedicated records grow, patterns slowly unveil themselves. Nobody ever told me this directly — I had to learn it.

You might discover that particular plants extend their blooms in response to increasing pollinator visits or notice a burgeoning population of ladybird beetles coinciding with a decline in aphids. Such insights fuel proactive gardening practices, allowing you to experiment with plant combinations, adjust maintenance protocols, and potentially introduce beneficial species to bolster ecosystem services. Let me tell you, this changed how I think.

As strange as it sounds, But perhaps most transformative in the act of monitoring and documenting is the way it reshapes our perception of time and change. It might sound dramatic, but this matters.

Gardens remind us of the cyclical nature of life, the dance of death and rebirth that every season exemplifies. As you chart your garden's journey through the years, you witness its evolution—not only physically, as saplings mature into trees and annual blooms paint new canvases every spring, but also in the subtle shifting of ecological dynamics. Let me tell you, this changed how I think.

These pages of observations allow your garden to become a testament to resilience and adaptation. Nobody ever told me this directly — I had to learn it.

When faced with challenges like climate shifts or pest incursions, the detailed records of your garden's past serve as a map to navigate its future. This history provides a beacon of hope, illustrating nature's extraordinary ability to adapt and flourish even amidst the trials of change. Let me tell you, this changed how I think.

In sharing your observations, you further enrich both your understanding and that of others. Nobody ever told me this directly — I had to learn it.

Connecting with local gardening clubs, community science projects, or online forums transforms personal notes into collective knowledge. By contributing to citizen science initiatives, your data on pollinator trends and plant health may aid wider ecological studies, enhancing conservation efforts and fostering global biodiversity awareness. Let me tell you, this changed how I think.

In real life, Ultimately, the art of monitoring and documenting garden biodiversity is more than a scientific endeavor; it is an expression of care and curiosity. Here's the truth nobody talks about.

It cultivates a deeper bond between the gardener and the garden, inspiring a sense of stewardship that transcends mere horticultural practices. As you pause to see, listen, and learn from your garden, you unlock not just its secrets but a profound understanding that we are all part of an intricate web of life. Let me tell you, this changed how I think.

In this intricate dance of observation, documentation, and reflection, your garden becomes a living canvas. I used to completely misunderstand this.

Each day, it paints new colors, ties fresh threads, and tells evolving stories. By engaging in this mindful chronicle of life's richness, you embrace the garden as a lifelong companion, a teacher that nurtures not only plants but the soul itself. It might sound dramatic, but this matters.

Chapter 18: Addressing Challenges: Weather, Pests, and Diseases

In the grand tapestry of a flourishing garden, weather, pests, and diseases are the complex hues that add dimension and challenge. Let me tell you, this changed how I think.

For every gardener aspiring to create a haven for pollinators and nurture thriving plants, confronting these challenges isn't just a necessity; it is an opportunity to learn, adapt, and emerge more in tune with nature's rhythms. Let us embark on this empowering journey of addressing such adversities, armed with wisdom and resilience. Let me tell you, this changed how I think.

Navigating the Capricious Weather

At first glance, Weather, nature's most unpredictable muse, can both embellish and undermine the success of a garden. Let me tell you, this changed how I think.

From unseasonal frosts to weeks of relentless rain, each weather pattern requires a thoughtful response. I used to completely misunderstand this.

Begin by harnessing the potential of preventive care. Get intimate with your local climate; track patterns and seasonal forecasts which can offer clues to what's ahead. It might sound dramatic, but this matters.

Use this knowledge to select resilient plant varieties well-suited to your specific conditions. Hardy perennials or drought-resistant native species often withstand the trials of unexpected climate changes better than exotic imports. It might sound dramatic, but this matters.

For frost-prone areas, sowing in protected environments such as greenhouses or cold frames can safeguard young seedlings. Here's the truth nobody talks about.

Conversely, during heatwaves, a regimen of mulching can help in retaining soil moisture, while a shade cloth can mitigate the sun's intense gaze, protecting delicate blooms and enriching your garden's microclimate. Let me tell you, this changed how I think.

When faced with prolonged rainfall, ensure your garden has been designed with proper drainage in mind. Nobody ever told me this directly — I had to learn it.

Raised beds, gently sloping surfaces, and well-placed ditches can prevent waterlogging, averting root rot and fungal diseases that thrive in overly saturated soils. I used to completely misunderstand this.

The Dance with Pests

In the enchanting world of your garden, pests may enter like uninvited guests at a feast. Yet, the presence of insects is an integral component of a balanced ecosystem. Let me tell you, this changed how I think.

Many of them, like the lowly aphid, are vital food sources for beneficial insects such as ladybugs or lacewings. Understanding this dynamic builds the stage for a harmonious dance of predator and prey. Here's the truth nobody talks about.

It's worth noting that Embrace an integrated pest management approach, which combines biological, cultural, and mechanical strategies, minimizing chemical interventions wherever feasible. It might sound dramatic, but this matters.

Planting a diverse array of herbs like dill, basil, and mint can attract beneficial insects or deter specific pests through their aromatic qualities. Here's the truth nobody talks about.

Consider physical barriers such as floating row covers which can shield seedlings from larger predators. It might sound dramatic, but this matters.

If manual intervention is necessary, handpicking larger pests like slugs or caterpillars, or using targeted solutions such as neem oil, can provide relief without compromising pollinator-friendly nooks of your eden. I used to completely misunderstand this.

Encourage natural allies by creating habitats that foster a thriving population of beneficial insects and birds. I used to completely misunderstand this.

Hedgerows, native plants, and water features can transform your garden into a welcoming sanctuary for these natural defenders. I used to completely misunderstand this.

Outwitting Plant Diseases

Diseases present a more subtle adversary, often disguising themselves until they've woven their way into the garden's lush tapestry. Here's the truth nobody talks about.

The key to curbing plant ailments lies in vigilance and proactive care. Here's the truth nobody talks about.

Start with the soil—it's the canvas upon which your garden masterpiece is painted. Nobody ever told me this directly — I had to learn it.

Ensure it is rich in organic matter and well-aerated to support robust root systems, which are your plants' first line of defense. Rotate crops annually to prevent the buildup of soil-borne diseases. Here's the truth nobody talks about.

Let's not forget, Keep a watchful eye on air circulation within your plantings; overcrowding encourages fungal growth by trapping moisture around leaves and stems. Here's the truth nobody talks about.

Pruning away dead or diseased growth promptly can help prevent the spread and instill vigor in your plants. Nobody ever told me this directly — I had to learn it.

When fungal infections emerge, such as mildew or rust, early intervention is crucial. Nobody ever told me this directly — I had to learn it.

Homemade remedies, including sprays made from baking soda or milk, can provide effective organic alternatives for disease management. Always aim to treat issues promptly, keeping records to learn from each encounter. Nobody ever told me this directly — I had to learn it.

Cultivating Resilience and Reflecting on Growth

Believe it or not, Every challenge faced in the garden is a lesson in resilience and patience. Let me tell you, this changed how I think.

A gardener's journey with weather, pests, and diseases parallels the cycles of life itself; it is a commitment to nurturing growth against odds, and an unwavering desire to learn from each setback. Here's the truth nobody talks about.

Documenting your garden's journey can be a powerful tool. It might sound dramatic, but this matters.

Keep a journal of seasonal observations, pest patterns, and interventions that worked or faltered. These insights, gathered over time, will serve as a personalized guide, refining your intuition and deepening your connection with the landscape you steward. I used to completely misunderstand this.

As you address these challenges, remember, you are part of a larger community of gardeners who have faced similar trials for centuries. I used to completely misunderstand this.

Tap into this vast network by sharing experiences, seeking advice, and collaborating on sustainable solutions that transcend individual gardens. Here's the truth nobody talks about.

At first glance, In each season, as you stand amid blooming flora, inhale the sweet fragrance of success that only comes after confronting adversity. Nobody ever told me this directly — I had to learn it.

Weather, pests, and diseases, while often daunting, are transformative maestros in the garden's symphony, teaching you to see beauty in imperfection and strength in vulnerability. Let them guide you to unlock your garden's fullest potential, as you grow alongside it, nurturing a world that welcomes pollinators and fosters life. It might sound dramatic, but this matters.

Chapter 19: The Impact of Climate Change on Pollinators

As gardeners, we harness the earth's natural rhythms, translating its whispering secrets into vibrant ecosystems that buzz with life. In these verdant sanctuaries, pollinators play a vital role, rendering them not just guests but indispensable agents in the grand tapestry of growth. Here's the truth nobody talks about.

However, like delicate threads woven too tightly, the effects of our changing climate are pulling at the seams, placing these tireless workers under unprecedented stress. Let us explore this intricate relationship, unraveling not only the challenges but discovering pathways of resilience and adaptation. Here's the truth nobody talks about.

The Shifting Climate Mosaic

It's worth noting that Our climate is not a static monolith; it is a complex mosaic of interdependent systems. As temperatures climb and weather patterns shift, we witness changes that echo through every layer of the biosphere. Let me tell you, this changed how I think.

For pollinators, these shifts translate into significant consequences. Temperature increases can disrupt flowering times, leading to mismatches between when plants bloom and when pollinators are active. This phenological decoupling can result in reduced food availability, and in turn, impact reproductive cycles. I used to completely misunderstand this.

Global warming also extends growing seasons in some regions—potentially a boon for gardeners but a conundrum for pollinators who rely on cues from nature. Here's the truth nobody talks about.

Earlier springs and more erratic weather risk throwing these well-honed instincts into disarray. This change challenges them to either adapt to new flowering schedules or face starvation. It might sound dramatic, but this matters.

It is here that the strength of our interconnectedness is tested, urging gardeners to rise as guardians of their local ecosystems. It might sound dramatic, but this matters.

Our decisions ripple outward, providing sustenance or scarcity to those fragile wings that flit among our flora. Nobody ever told me this directly — I had to learn it.

It's worth noting that #### The Stressors of Temperature and Extreme Weather

Pollinators are remarkably sensitive to temperature fluctuations. Bee species, crucial for pollinating oilseed, fruits, and vegetables, are particularly vulnerable.

Let me tell you, this changed how I think.

A mild rise in temperature can lead to increased metabolic rates, making it essential for bees to forage more frequently to sustain their energy levels. With erratic weather, a reliable food source becomes uncertain, causing severe stress on the colony's survival. I used to completely misunderstand this.

Extreme weather events—those torrential floods, prolonged droughts, and unseasonable frosts—compound these challenges by destroying nests and reducing the availability of wildflowers and other natural food sources. It might sound dramatic, but this matters.

For instance, heavy rains can wash away the delicate, time-honored nests of ground-nesting solitary bees, while droughts wither the once multicolored quilt of wildflowers into a barren patchwork. Here's the truth nobody talks about.

I've seen this happen before: As guardians of the garden, we can offer a stable refuge amid nature's growing volatility. Nobody ever told me this directly — I had to learn it.

By selecting a diversity of native plants that stagger their blooming across months, we ensure a continuous nectar supply, reducing the threat posed by brief shortages. Installing bee hotels and offering encouragement to solitary pollinators through deliberately left patches of untouched soil can add threads of resilience to the ecological tapestry. Nobody ever told me this directly — I had to learn it.

Rising Carbon Dioxide and Pollen Quality

The increase of carbon dioxide levels in our atmosphere has a less visible, yet equally profound effect on our garden's ecosystem. Scientific studies suggest that heightened CO2 can alter the nutritional quality of pollen. Here's the truth nobody talks about.

Plants like ragweed have shown increased pollen production under higher CO2 conditions, yet with a noticeable decrease in protein content. Protein is critical for pollinator development, especially for rapidly growing bee larvae. Here's the truth nobody talks about.

To be honest, Imagine your garden's pollinators dining at a sumptuous feast, only to find the meals increasingly devoid of the nutritional value that sustains life. Our charge, then, becomes to enhance their diet actively. It might sound dramatic, but this matters.

Plant a variety of flowers with known nutritive benefits, like buckwheat and canola, which offer rich, protein-filled pollen. The careful selection of plant species can make a tangible difference, evoking nature's abundance even in times of paucity. Nobody ever told me this directly — I had to learn it.

Migration and the Struggle for Coexistence

Climate change nudges the borders of pollinators' habitats, often pushing species to migrate to cooler climates or higher elevations. Nobody ever told me this directly — I had to learn it.

This migration can lead to competition between native species and those forced to move, creating a struggle for resources in already fragile ecosystems. Here's the truth nobody talks about.

Many people don't realize this, but To cultivate coexistence, let us embrace a new vision for our gardens—a vision that promotes inclusivity and harmonizes with these shifting tides. It might sound dramatic, but this matters.

By focusing on planting a wide range of native and climate-resilient flora, we offer an olive branch to both transient and resident pollinators, contributing to the thriving diversity necessary for a healthy garden ecosystem. I used to completely misunderstand this.

Cultivating Hope and Harmony

In the face of this challenge, let our efforts not wane into despair, but swell into a tide of possibility—a renewed dedication to our shared earth and the marvelous creatures that animate it. Here's the truth nobody talks about.

Simple actions, like selecting the right plants, creating layered plant structures, and setting aside wild areas, weave hope into the garden's fabric. Nobody ever told me this directly — I had to learn it.

At first glance, Education becomes our ally. It might sound dramatic, but this matters.

By understanding the dynamics of climate change and its impact on pollinators, our actions can inspire our communities, sparking a collective consciousness and drive to protect our vital allies. Nobody ever told me this directly — I had to learn it.

In this grand narrative, your garden is an epicenter of change, where every seed sown and habitat restored speaks of an enduring legacy, a commitment to a future enriched by vibrant networks of life. Let me tell you, this changed how I think.

Together, let us draw inspiration from the resilience of nature, cultivating gardens that stand as testaments to creativity, adaptation, and harmony. Nobody ever told me this directly — I had to learn it.

Even as climate's uncertain brush strokes color the world anew, we remain architects of landscapes where pollinators can thrive, guiding nature toward a symphony of life where every voice is cherished, and every flight is celebrated. Nobody ever told me this directly — I had to learn it.

Chapter 20: Cultivating a Legacy: Sharing Knowledge and Inspiring Others

As you stand amidst the vibrant blooms and hear the hum of busy bees in your garden, you're not just a gardener; you're a steward of nature's abundant legacy. Your garden is more than an array of plants and flowers—it is a living classroom, a source of inspiration and education that has the power to impact future generations. Nobody ever told me this directly — I had to learn it.

Cultivating this legacy involves not only growing plants that attract pollinators and encourage growth but also sharing your knowledge and passion with others. This chapter will explore ways in which you can reach beyond the garden gate, inspiring others to create and nurture their own green spaces. I used to completely misunderstand this.

The first step in sharing your knowledge is storytelling. Every gardener has a story to tell. Whether it's the tale of triumph over a challenging soil condition or the joy of seeing the first butterfly of spring, these narratives have the power to ignite curiosity and passion in others. Here's the truth nobody talks about.

Storytelling humanizes the science of gardening, transforming abstract concepts into relatable experiences. Share your gardening stories through social media, blog posts, or local community newsletters. Engage with your audience by showing them the wonders of your journey—the successes, the failures, and the lessons learned. By making your experiences accessible, you open a gateway for others to embark on their gardening adventures. Nobody ever told me this directly — I had to learn it.

Believe it or not, Workshops and demonstrations are another practical method for sharing your gardening expertise. Organizing a workshop can be as formal or informal as you like, ranging from a structured course at a local community center to an informal garden tour in your own backyard. Nobody ever told me this directly — I had to learn it.

Consider focusing on topics that highlight your strengths, such as organic gardening techniques, constructing bee hotels, or companion planting for pollinators. Encourage participants to get their hands dirty, experiencing the tactile joys of gardening themselves. Hands-on learning not only enhances understanding but also creates lasting impressions, allowing participants to walk away with both knowledge and enthusiasm. It might sound dramatic, but this matters.

Another route of inspiring others is through mentorship. Consider reaching out to neighbors, schools, or youth organizations to offer guidance to aspiring gardeners. Here's the truth nobody talks about.

Mentorship provides a unique opportunity to foster a personal connection and offer tailored advice. This one-on-one interaction allows for a deeper exchange of knowledge and can be incredibly rewarding, both for the mentor and the mentee. You'll have the chance to witness someone else's garden flourish under your guidance, contributing to a ripple effect of growth and inspiration. Here's the truth nobody talks about.

Furthermore, hosting community gardening projects can be a remarkable way to engage others, create a collaborative atmosphere, and beautify a shared space. Approach local governments, schools, or neighborhood associations with the idea of establishing a community garden. Here's the truth nobody talks about.

With a collective effort, these projects not only promote environmental health but also build social ties and a sense of ownership among participants. People come together, share resources, and learn from each other. In a world that can sometimes feel disconnected, community gardens reconnect us to the earth and each other, fostering a sense of unity and purpose. Here's the truth nobody talks about.

Many people don't realize this, but Utilizing the internet as a tool for education and inspiration is vital in today's digital age. Create online content that captures the essence of your garden and translates it into digestible, shareable information. Nobody ever told me this directly — I had to learn it.

This could be in the form of a video series explaining plant care, a podcast discussing the benefits of pollinators, or a virtual garden tour showcasing your hard work. Online platforms broaden your audience far beyond your local community, allowing you to reach and inspire a global readership. Collaborate with like-minded individuals and platforms to expand your influence and share fresh perspectives from others in the gardening community. Let me tell you, this changed how I think.

And let us not overlook the power of writing. If you're inclined towards the literary, consider writing articles or even a book on your gardening experiences and insights. It might sound dramatic, but this matters.

Contribute to gardening magazines, or self-publish a collection of essays. Writing demands reflection and articulation, deepening your own understanding while nurturing others. Through the written word, your garden's legacy can reach countless readers, sowing seeds of inspiration across time and place. Here's the truth nobody talks about.

Through each of these activities, emphasize sustainability and symbiosis as foundational principles. Here's the truth nobody talks about.

Show how small acts, like planting native flowers or avoiding chemical pesticides, contribute significantly to local biodiversity and environmental health. Inspire others to view their gardens as a form of environmental activism—an opportunity to make a tangible positive impact on the world around them. Here's the truth nobody talks about.

At first glance, Ultimately, the most effective tool you possess in cultivating a legacy is passion. Genuine enthusiasm is contagious, and when others observe your love for gardening, they are more likely to engage with their own natural surroundings. Here's the truth nobody talks about.

Be encouraging and inclusive, fostering a community where everyone—regardless of experience level—feels welcome and valued. Diversity in knowledge and approach is as beneficial in the gardening community as it is in nature, strengthening the movement to create pollinator-friendly environments. Let me tell you, this changed how I think.

As your journey from gardener to guardian unfolds, remember that you are integral to a grander narrative. Standing at the intersection of nature and nurture, your influence ripples outward, touching each person you engage, each garden you shape, and each seed you sow. I used to completely misunderstand this.

The legacy you cultivate transcends the physical act of gardening; it plants ideas, grows communities, and pollinates the minds of future caretakers of the earth. As you continue to share your knowledge and inspire others, know that you are sowing the seeds of resilience, hope, and positive change—ensuring that the garden of life thrives for generations to come. I used to completely misunderstand this.

About the Author

T. S. AVINI is a writer who believes in the power of sharing life's lessons. Having faced hardship and pain, he turned to writing as a form of self-expression and healing. Today, he pours his heart into every book, aiming to help others navigate their own struggles and find strength in vulnerability. Let me tell you, this changed how I think.